THE SPACE HERO'S GUIDE TO GLORY

HOW TO GET OFF YOUR PODUNK PLANET & MASTER THE FINAL FRONTIER

NICK HURWITCH & PHIL HORNSHAW

Published by Sourcebooks, Inc.
P.O. Box 4410, Naperville, Illinois 60567-4410
(630) 961-3900
Fax: (630) 961-2168
www.sourcebooks.com

Library of Congress Cataloging-in-Publication Data

Hurwitch, Nick.
  The space hero's guide to glory : how to get off your podunk planet and master the
final frontier / Nick Hurwitch & Phil Hornshaw.
    pages cm
 (trade : alk. paper) 1. Space flight—Humor. 2. Outer space—Humor. I. Hornshaw,
Phil. II. Title.
  PN6231.S645H88 2015
  818'.602—dc23

2014032970

Printed and bound in the United States of America.
VP 10 9 8 7 6 5 4 3 2 1

CONTENTS

× THE ×

SPACE HERO

★ ★ ★ ★ ★ ★ ★ ★ ★ ★ ★

HALL OF GLORY

AND

GIFT CATALOG

This book pays homage to the many Space Heroes who have blazed interstellar trails to make Zero G adventure possible for those who came after. You can pay homage to them and support the Spacefleet Operational Service by ordering these limited-edition gifts for the prospective Space cadet in your life!

Captain James T. Kirk, USS *Enterprise* (NCC-1701-A)
Commemorative Space Chess Set (*Now with real holograms!*): 1,500 credits

Captain Han Solo, Rebel Alliance, *Millennium Falcon*
Collectible "ALWAYS SHOOT FIRST" shooter shot glass (*The Holy Grail of shot glasses*): 300 credits

Captain Malcolm "Mal" Reynolds, *Serenity*
Rare "Smugglin'" Crate (*It's full—of potential!*): 1,400 credits

Commander Luke Skywalker, Rebel Alliance
Official Jedi Order "My Master and Me" Backpack Training Harness (*available in Tatooine Tan and Dagobah Swamp Green*): 1,200 credits

Captain Kathryn Janeway, USS *Voyager* (NCC-74656)
"My First Borg" USB-powered Pint-sized Cybernetic Organism (*Resistance to cuteness is futile!*): 450 credits

Captain Jean-Luc Picard, USS *Enterprise* (NCC-1701-D)
Limited Edition Skin-toned Skullcap (*Not bald? Make it so!*): 600 credits

Commander William "Bill" Adama, Battlestar *Galactica*
Replica Caprican Law Books, Complete Set (*LEGALLY outdated*): 2,000 credits

Lieutenant Ellen Ripley, USCSS *Nostromo*
Tabby Cat "Jonez" Clone (*Meow*): 750 credits

Commander Ender Wiggin, International Fleet (IF)
Alien Murder Simulation Video Game (*It's totally not real!*): 500 credits

Captain Kara "Starbuck" Thrace, Battlestar *Galactica*; Viper 8757
Cigar Sampler Gift Basket with "Galactica Top Gun" Commemorative Stein (*You'll be frakking in no time*): 800 credits

Commander John Crichton, IASA
Hand-crafted John Crichton Hand Puppet (*Command the Commander!*): 1,000 credits

Captain William Anthony "Buck" Rogers, NASA, *Rebel 3; Searcher*
"GREETINGS FROM THE TWENTY-FIFTH CENTURY" vintage postcard (*What's a postcard?*): 50 credits

Barbarella, Queen of the Galaxy
Queen of the Galaxy brand hair spray, 58 oz. (*Guaranteed to hold in zero to four g*): 669 credits

General Lando Calrissian, Rebel Alliance, *Millennium Falcon*
Foil-card Sabacc Deck with Velvet Carrying Case: 750 credits

Commander Shepard, Systems Alliance, SSV *Normandy*
Impact-Proof Systems Alliance Starship Aquarium (*200 gallons; fish not included*): 27,000 credits

CAPTAIN'S LOG

Stardate...Thursday, if memory serves. If you're reading this, you should know that the following transcription will not be for the faint of heart. Death off the port bow! Regret off the aft! This will be...my final captain's log.

Long and battle-hardened though my log has become, it shall soon be freed from the cramped synthetic confines of the SOS *Starhawk Flamepanther* mainframe and, from there, explode into the same illustrious renown as its author.

But who is this author? you're surely asking yourself. *Why does his voice sound so sensual as I imagine it ringing within my skull? Could I—a lowly ensign or simplistic terrestrial citizen—manage to arouse myself as quickly and effortlessly as he has just done?*

The answers to which, in reverse order, are: probably not; merely one of the Universe's great mysteries; and Dirk Parsec, the Universe's finest captain, smuggler, lover, shot, pilot, diplomat, hero, and, if you believe the rumors, singer-songwriter. Leader of brave men, women, robots, and aliens alike. Or at least I *was* a leader of brave men, women, and robots. They're all evacuated now. Back to the stars.

As I am also a gifted xenolinguist, let me paint for you a word picture: the retina-searing beauty of a cerulean star overwhelming the faint ruby emergency lights of the *Flamepanther*, pride of Sector 18. In a matter of hours, the ship and I will hurtle into the bright blue star out our view port, though I'll be dead long before. Suffocated, most likely, by lack of oxygen. Or perhaps incinerated from unfathomable heat, the *Flamepanther* my makeshift urn.

Yes, this is the end. The last hurrah. The final engagement. The sign-off shindig. The quintessential quietus. The red-carpet stroll into Satan's foyer. Even as I speak, the *Flamepanther* and I are being pulled into the star's gravity, well past the event horizon.

So before I die, I thought it prudent to write down by way of automated dictation all that I know about Space and Space Heroism. To pass along my seemingly impossible wealth of knowledge, in hopes that some youngster out there in the cosmos might one day take up the mantle of Greatest Spacefaring Captain in the 'Verse. Someone with the wherewithal and proverbial gonads to reach out into the ether and pull back that which is necessary to forge the next generation's great Space Hero.

The odds that this means *you*, specifically, are astronomically low—even without taking into account the statistical improbability of 1) finishing the book before I perish; 2) this book finding its way to intelligent life; or 3) this book being published and 4) it becoming an intergalactic bestseller, thereby adding trillions to my estate posthumously.

But my time has come. And so has yours! We do not choose our deaths in the black vastness of infinity. We can only aim to dodge our demise for as long as possible before we gaze upon the Grim Reaper's Space scythe at last. Nor do we choose our lives, tugged along as we are by the inertia of infinity...But we can choose to shape them!

Let's hear it, then. Do you want to be *you*, entropy taking hold of your pathetic appendages as you waste away in your relaxed position of choice? Or do you want to be a Conqueror of Space! A Smuggler of Fine Wares! A Pleasurer of Alien Babes! A Champion of the People! A *Space Hero*!

Set your phaser to *stunning*. Space needs you.

Captain Dirk Parsec

EPISODE I

A NEW HERO

I believe every human has a finite number of heart-beats. I don't intend to waste any of mine.

—Neil "Erroneously Attributed" Armstrong

CHAPTER 1

THE HERO YOU'LL BECOME

The first step to becoming a Space Hero—beyond getting into Space and not immediately falling into a black hole or having a control panel explode in your face—is to distinguish yourself from the legions of technicians, marines, scientists, pilots, navigators, xenobiologists, regular biologists, doctors, engineers, astronomers, botanists, janitors, cooks, bartenders, pirates, monks, religious zealots, and colonists floating around out there with you.

But much of what is required can't be taught. You'll need guts, gumption, and force of will. You'll need the passion to take on seemingly impossible tasks and the style to make those tasks look easier than ignoring a protocol droid. You'll need more lives than a twice-cloned cat and a one-liner in your pocket for every vile situation—from encountering disgusting new alien species to discovering the recently mutilated corpses of crewmates.

There are a number of great Space folk out there. But to be a true Space Hero, you need to be more than an interesting person who merely happens to be in Space. You need to become an interesting *protagonist* who merely happens to be in Space.

Here's a brief list of qualities the Spacefleet Operational

Service (SOS) has identified that you need to possess, obtain, or at the very least emulate. All are important if you wish to become a Space Hero—and some are necessary just to become a useful Space janitor. Consider this your baseline before even thinking of venturing into the void.

(1) **The ability to stare into the yawning jaws of death and calmly pick the remains of your friends from its teeth.** Space is dangerous. You need to accept that and, in time, be unfazed by it. After all, you'll be dealing with the horrors of radiation sickness, Space madness, vacuum exposure, and muscle atrophy every day for the rest of your magnificent life. It's unbecoming of a Space Hero to fall to pieces every time an engineer needlessly gets his head lopped off by a dangerous piece of machinery during a routine meteor shower. Who cares if he was a galaxy-class masseuse or the last of his species or whatever. Show some composure.

(2) **A vast working knowledge of Science.** Your mother and father and Guardian Robot Television were right: school was not just about sexing up the Prom Pansexual Monarch, but about learning. Whether you are the next Space Hero or one of thousands of functionaries who will do the bidding of the next Space Hero, you'll need to possess a functional heap of gray matter.[1] You can only skate by on your wits and your devilish good looks for so long. Unfortunately, there's more to the life of a Space Hero than making important decisions, outsmarting enemies, and drinking the finest Space scotch. Like years of school. Hope you like Science.

---

1. Not to be confused with dark matter, of which your brain should *not* be composed.

The kinds of people who live and work in Space spend years training for it. They are in peak physical and mental condition. They possess intimate knowledge of the many complex systems required to stay alive in vacuous, unforgiving infinity. And they're capable of carrying out orders, regardless of how insanely complex or so-crazy-they-just-might-work those orders may be. You may be asked to attempt a Malcolm Reynolds diet (dumping your valuable cargo of smuggled goods for an extra boost of getaway speed) or perform an Adama maneuver (hurtling toward a planet and warping out of harm's way just before crashing into the surface).

With any luck, you may be the one to *give* such an order. If you think you can stand in the shoes of the greatest of Space captains in Space history without knowing about relativity, xenobiology, matter-antimatter fuels, and the psychology of Klingon command structures, you're sorely mistaken. You'll get through that and more in your Monday morning briefing alone.

You might not be a prodigy. You likely won't start at the top. But it's never too late fill your gray matter with Space matters.

(3) **The willingness to sacrifice yourself for the good of your captain (and others).** Space is a community. We all work together for the benefit of our fellow Space(wo)man. Of course, some Space(wo)men are more important than others. There may come a time when you'll be asked to make the ultimate sacrifice for the good of those who bear a greater importance to the mission at hand or are more adored by the public. Even if you are a Space Hero, you, too, might one day ride your captain's chair into

the nuclear center of a blue dwarf, like many a captain before you.

Much more than likely, someone will be needed to stick a thumb in a hull breach or be the sacrificial bargaining chip that sates a hungry alien race during diplomatic negotiations. Living in Space is fraught with risk, and the needs of the many outweigh the needs of you.

(4) **Rugged good looks.** Even in a utopian Space society, the sad fact is that the ugly people are deckhands, and the beautiful people are sent on first-contact missions. It may not be fair, but as it turns out, most aliens prefer a heaving bust and throbbing pants bulge. Sometimes on the same person. If you're a handsome specimen, people will presume you're smart, assume you're telling the truth, and take for granted the fact that your beauty may be allowing you to swindle them, even as they shake your firm, well-manicured hand. But even if your face has more craters than a rogue asteroid named Bill Adama, make your imperfections count for something. Consider how much *more* badass Luke Skywalker became after getting his face jacked by an ice-cave-dwelling wampa (alliances with primitive teddy bear warriors notwithstanding).

THE MANY PATHS TO SPACE HEROISM

Part of the point of becoming a Space Hero is to be known by all: admired by men and women, adored by various alien love-slaves, and glorified by children and forest creatures. What's the point of dying heroically if everyone (anyone) forgets about you afterward? You must strive to be not only memorable, but unique (unless you clone yourself).

So even though your deepest and most primal desire is to be just like Kirk or Kara or Han, you must strive to be your own hero. The tools are already within you: the traits innate, genetic, or born from your (soon-to-be) tragic backstory (which we'll explore in Chapter 2).

Despite the requirement to be singular, there are Spaceman archetypes: categories into which you'll find yourself stuffed like a long-sleeved swindler in a Nar Shaddaa prison cell after an evening of high-stakes mah-jongg gone south. If one of the following heroes appeals to you, there's a chance you'll be able to make a push toward your preference. Or, if you already see yourself reflected in his or her stark and stunning features, prepare yourself. That could be your face on the cover of the twenty-eighth edition of this book.

THE SPACE MONKEY HERO

Examples: Neil Armstrong, whatever that Space monkey's name was, Ender Wiggin, Ellen Ripley, Laika the Space dog, Dirk Parsec

A Space Hero secret: if you are the first to do something, you automatically get to be lauded as a hero. You'll be a pioneer of Space discovery! Even—perhaps especially—if the thing you've done hadn't been done before because of the seeming likelihood of death and perceived degree of stupidity required. The first monkey shot into Space in 1948 died horrifically and against his will. But he was thusly rewarded with the title of Hero in the Great Beyond.

Should you do something as first and as stupid as: launch yourself out of a torpedo tube because your ship is out of ammunition (Valentina Rahimovic); have intercourse with a betentacled alien life form (Hugh Huffer); set foot on a previously unexplored celestial body (Neil Armstrong); or discover the Pant-Tightness-to-Character Importance Ratio (Captain James T. Kirk), you too could back your way into hero-dom, regardless of what you accomplish henceforth.

While this might seem like your easiest path to Space fame, an increase in fame-seeking opportunists like yourself and a decrease in things that have never been done before mean that the chances of becoming a Space Monkey Hero are dwindling. Additionally, if you die, you may be reviled as an idiot instead of heralded as a pioneer, or end up with a small statue or a postage stamp as the only evidence of your greatness.[2] But if you *don't* die, it's possible no one will care. There's really no way of knowing before you try.

THE WAR HERO

Examples: Commander Shepard, Benjamin Sisko, Malcolm Reynolds, Ender Wiggin, Leia Organa, William Adama, Dirk Parsec

This one requires patience, dedication, and a large dose of galactic intervention. You can be the greatest soldier in the 'Verse, train diligently, climb the ranks, strike fear into the hearts or comparable ventricle sacks of alien scum from here to the Crab Nebula…but if there's no war, you won't even make the Hero-meter shiver.

If this is your path, make sure there's a war on. Or make sure to provoke a war into being on.

---

2. In fact, Laika the Space dog has both a statue and several postage stamps in her honor, which is pretty good for a dog, although this has been cited as an example of anti-monkey discrimination. Stamps featuring monkeys as of 2399: zero.

But tread as lightly as your artificial gravity systems will allow. There are no shortcuts to becoming a hero of war. You'll have to save your fellow soldiers, brazenly sacrifice yourself for the greater good while also surviving said sacrifice, and wear a really heavy Space suit at all times, even while sleeping or pooping or both. You will sweat your glands out and then be forced to drink that sweat to stay alive.

You'll also have to be mildly insubordinate and covertly opportunist. The difference between a soldier with commendations and a fabled war hero is typically a combination of daring and stupidity, but the reward is often a command and a captain's chair. If heroism is what you're after, you'll have to steal it outright or earn it through sheer force of harebrained will. No one ever blew up a Death Star using a targeting computer, after all.[3]

STAY STUNNING! WITH DIRK PARSEC

Securing Your Legacy

There's a chance that whispers of a war will make their way through your nameless, podunk town; that the enemy-army-to-be will incinerate that nameless, podunk town; that you'll swear revenge, enlist, become the ultimate badass; and...that the two

---

3. Another good tip for surviving a war: don't be named Porkins. If your commanding officers can't say your name without giggling, there's no way you're going to survive that Death Star trench run—I don't care how good a pilot you are.

sides will come to a peaceful accord. But this does not mean your path to heroism has been cauterized. Nay! It has merely been circumcised. You can go rogue. Expose the dark underbelly of your lying government. Become a smuggler. As long as you're principled and honorable within reason, a military defection doesn't have to mean heroic dysfunction.

THE BASTARD ANTIHERO

Examples: Han Solo, Malcolm Reynolds, Kara "Starbuck" Thrace, Dirk Parsec

If you were the type of kid who sat in the back of class flinging spitballs, sizing up the teacher's bust or nondescript bulge, and keeping mostly to yourself, you might be a bastard. But if you were also the type of kid to step in and throw the first punch when the nerd was getting picked on...you might be a bastard antihero.

You have a chip on your shoulder and a gun on your hip. You hold your love of "the little guy" behind an icy, selfish exterior. Personal failures (and a tragic past) have made you disenchanted in your relationships with establishments, governments, groups, and society at large. A past betrayal has turned your heart to stone and made you reluctant to get too close to anyone but a trusted one or two.

You live by your own moral code, which is not so much a code, but a loosely packaged collection of "feelings" revolving around self-preservation that express themselves as needed. And just when this description is getting too touchy-feely, you also like to shoot those fools who get in your way—first, often, and occasionally while they're surrendering to you. You enjoy gunning down your enemies almost as much as delivering a catchy one-liner as a preamble, much to the delight of your ragtag troop of loyal followers.

Just don't forget to stress the "hero" part as well as the "bastard" part—sure, you might lie, cheat, and steal, but you do so for the good of your team and with often-honorable intentions. As long as you don't become a sociopath and generally shoot only the folks who deserve it, you'll be right at home here. Plus, you get to do illegal crap with no (some) moral ambiguity. Sticking it to the Man is your whole shtick!

THE HERO OF DESTINY

Examples: Luke Skywalker, Anthony "Buck" Rogers, Kara "Starbuck" Thrace, Dirk Parsec

As though you need a reminder, there is nothing remarkable about you. Before saving the galaxy from certain doom, no one could have guessed that a character as anonymous and poorly dressed as your-self could amount to anything more than one day inheriting your aunt and uncle's moisture farm and taking a SexBot for a spouse.

Yet you have something they cannot see. No, not midichlorians—destiny! That's right. There may be more than one powerful force flapping around under your weird desert dress. And if there is, you belong to the most exclusive club of heroes, whose members' nondescript personalities only serve to bolster their meteoric rise to interstellar stardom. The more dramatic the arc, the more impressive the hero.

It won't always be easy for you. Little about your monotonous farmhand life could possibly prepare you for the twists and turns that lie ahead. You may discover that you're adopted or the last of a dying breed of Space monk or are sexually attracted to your twin sibling. But it is no matter: destiny has the helm! You need only ride where she takes you. And should destiny ever stick a galactic-sized Space fork in your path, you need only ask yourself: What would Dirk Parsec do?

THE ROGUISH LEADER HERO

Examples: James T. Kirk, Jean-Luc Picard, Leia Organa, Kathryn Janeway, John Crichton, Dirk Parsec

An upbringing in the Star Navy Academy or the rough equivalent for your colony or species, an incredible tactical sense, a renegade attitude, and an unwillingness to leave any comrade behind—that's what it takes to be the kind of roguish leader that inspires your troops to throw themselves willingly into the jaws of death for your amusement.

This is something of a late-development hero role; you'll need to prove yourself a leader and then lead people, and then not get them all killed on your first mission. But if you can manage some incredible tactical brinkmanship and shed a tear for any crewmates who don't make it back, you may find that you're a roguish leader.

The key here is to be an unrelenting do-gooder. You want the people working with you to love the fact that you're willing to do anything for them and your enemies to be annoyed with your supersappy, no-man-left-behind attitude. And then you hit them with the roguish part of your heroism by firing proton torpedoes up their asses just when they think you're too weak to sacrifice the lovable robot member of your bridge crew. Extra points if you manage to save the robot's head or memory banks for an emotional Space burial.

THE SPACE HERO'S WORKBOOK:

What Type of Space Hero Are You?

CHAPTER 2

YOUR TRAGIC ORIGIN STORY

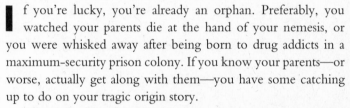

I f you're lucky, you're already an orphan. Preferably, you watched your parents die at the hand of your nemesis, or you were whisked away after being born to drug addicts in a maximum-security prison colony. If you know your parents—or worse, actually get along with them—you have some catching up to do on your tragic origin story.

The well-adjusted son or daughter of pioneering Space colonists? Bor-ing! The well-educated spawn of academy professors, raised and cared for as much by them as their enthusiastic students? Pass. The child of rich diplomats who traveled the galaxy and experienced all its exotic delights? Come on, people!

The only circumstances under which a coddled youth makes an acceptable foundation for heroism involve that youth being cut short with equal and opposite tragedy. You must use the burning corpses of your caregivers as fuel! Not literally...unless, of course, it is necessary for you to survive long enough to exact your revenge.

No sorrow in your spit-polished adolescence? There is still hope for you. You can probably come up with a reasonable facsimile of unfathomable tragedy, even if your immediate family remains less than crispy.

In any case, you surely have *something* you can spin into a tragedy. A tragically misplaced poodle. A tragically scuffed pair of dress shoes. A tragically inept round of school-yard fighting in which you had your ass tragically handed to you. Anything will work. You just have to *believe* it. And more importantly, the people you tell it to have to believe it.

A KNOWLEDGEABLE AND INSIGHTFUL MENTOR

Nobody becomes a Space Hero on his or her own. If you go it alone, you face a fate worse than death: no one mourning you after you die. And because no one showed you the ropes, you also face, you know, the death part. What you need is a mentor.

A mentor can be anyone—but is probably that hermit who lives on the outskirts of town. Provided he's not just homeless or insane, that hermit is the vehicle of your destiny. The *other* vehicle of your destiny. One that will guide your behavior, push you toward the unthinkable, and motivate you to see your destiny through after he or she inevitably perishes.

Yes, your mentor will die. And, yes, the experience is likely to be a great deal more painful and meaningful than the death of your parents or legal guardians. Why are you still thinking about them, anyway? Pull yourself together! They've been dead for several hours.

GETTING INTO SPACE

Congratulations! If you've been following along, those you love most in the world are now dead. They have been summarily replaced by an old weirdo of questionable mental stability. The only life you have ever known is gone and can never be returned, no matter how much you plead with your archaic god or gods. You, young hero-to-be, are ready for Step 2!

Though the circumstances of finally leaving your home are a good deal more horrific than you ever imagined, the time has come nevertheless. You have many options, but your goal is singular: you're getting off your podunk planet…and into *Space*.

The fact that you're going is not up for debate. But on the matter of how, some choice exists:

1. ENLIST IN THE SPACE MILITARY

Any planet worth its weight in Bantha poodoo has, or has been taken over by, some form of Space military or government. Provided it's not a robot or clone army—instances in which your heroic good looks would disqualify you from service— you could be in orbit and dry heaving into your standard-issue Space helmet by the time you're out of this paragraph. All you have to do is sign up.

Even during peaceful times, new Space military recruits die at such an incredible rate during training missions and improperly equipped Space walks that there's likely not even a waiting list. Just stumble aboard the nearest military cruiser and worry about how you'll avoid execution for going AWOL at a later date.

⊕ ENLISTMENT PROS:

- ⊙ FREE
- ⊙ FREE gun
- ⊙ FREE helmet
- ⊙ FREE ~~human shields~~—er, fellow combatants

⊖ ENLISTMENT KHANS:

- ⊙ Conscription
- ⊙ Superior officers
- ⊙ FREE Space gruel
- ⊙ High plausibility of getting shot
- ⊙ High plausibility of Space madness, explosive decompression, or psychic torture by godlike aliens

You can find out more about military Space duties (and other career paths for future Space Heroes) in Chapter 5.

2. STOW AWAY ABOARD AN OUTBOUND VESSEL

Whether it's the military, a Space science agency studying meaningless things like the color of rocks on other planets, or an icy smuggler with a secret soft spot for plucky adventure-seekers like yourself, someone's exiting the orbit of your planet sometime soon. It might happen every few minutes or every few years. Either way, if you can't pick one of the other (much better) options for getting into Space, you could stow away on the next transport, provided the following:

- ⊙ You've observed the captain of your target vessel and

he-she-it doesn't seem like the type to launch you out the air lock upon discovery.

◉ You've found a compartment hidden enough that you won't be found until it's too late to take you back, but not so hidden that it won't receive any of the ship's precious resources, such as heat and air.

◉ You've scoped out the ship's crew and contents before-hand and know that the crew will eventually allow you to befriend them, or at least sustain you until they ditch you at the nearest abandoned moon colony. Ships rarely travel with more food and water than is absolutely necessary, so an extra mouth to feed can be the difference between life and death. Specifically, your death.

◉ If they seem unlikely to allow you to befriend them or at least share their ration bars, you may need to bring your own staying-alive resources or manage to pilfer what you need from other areas of the ship by sneaking through the air and floor ducts. Expect to need roughly a liter of water and 1,200 calories every day to maintain health during the trip. (You'll need more if it's especially hot or if you need to flee security bots or other hazards.) If you're talking sub-light-speed travel, you could be facing years of stowaway time between destinations—which means you'll need space to store thousands of liters of water and millions of calories of food for the whole trip. The crew certainly won't be saving any for you. Thus, stowing away for any length of time might not be such a good idea.

◉ You've made an honest assessment of what you're willing to offer in exchange for safe passage. This includes swabbing poop decks, the use of your hidden talents, and a willingness to be a test dummy for Space foods, Space plants, Space drugs, and Space pranks.

There's always a chance you'll never be found and make your passage from one nameless crap-planet to another, trapped in the cargo hold. It's a rather awful scenario—most cargo holds are cramped, not especially clean, and the notorious hangout of Space bugs. Keep your orifices covered (I'm not kidding about the Space bugs), and begin building a tolerance for the acrid stink of your own feces.

Luckily, when you finally arrive at your destination, you can always lump your awful cargo-hold experience into your tragic backstory—provided you survive.

⊕ STOWAWAY PROS:

- FREE (provided you're not found)
- Skip the whole "human interaction" part that comes with negotiating safe passage or being a crew member—at least until you're found
- Skip the whole "doing work" part of being a crew member—at least until you're found
- Minimal likelihood of being murdered along with the crew in worst-case scenarios

⊖ STOWAWAY KHANS:

Being found may result in a number of undesirable outcomes, including:

- Getting tossed out an air lock
- Getting recycled or eaten or both

- Getting sold into Space slavery
- Being forced to act as if the crew is not made up of dullards and drunks, and that you'd like to be friends
- Dishwasher duty

Not being found may also result in a number of undesirable outcomes, including:

- Freezing to death
- Starving to death
- Dehydrating to death
- Being eaten to death by Space bugs
- Space madness

3. Steal a Ship

If you don't feel you're yet prepared to navigate the choppy social waters of Space relationships, you'd best avoid voluntary military conscription or forcible insertion into the delicate balance of an ornery crew. For now, you'll go it alone.

Unfortunately, this does not spare you the necessity of boarding a ship. Instead, you'll get to the high road of Space by traversing the low road of crime. Stealing a ship requires similar reconnaissance to stowing away. The ship must be properly fueled, stocked, and crewed by men, women, and aliens unlikely to skin you alive when they inevitably catch up to you. The crew also can't be on board when you steal the ship. Otherwise, you'll have just become history's worst stowaway.

✚ Theft Pros:

- FREE ship

- No requirement to deal with rightful ship owners
- Smugness at a job well done
- Chance to stick it to no-goodniks or Space hippies from whom ship was stolen

⊖ THEFT KHANS:

- Attention of interstellar police force attracted
- New enemies in the form of the ship's rightful owners
- Temperamental air conditioner that never seems to work right, and you can't figure out how to fix it
- Unknown, remotely activated self-destruct devices

4. BOOK YOUR PASSAGE WITH SOMEONE DESPERATE

Life in Space can be trying, nearly as trying as calculating currency exchange rates from solar system to solar system. Which is why, no matter how unwelcoming they seem, most captains can be bought. Now that you've sold your SexBot, ditched your old speeder, or rented out the charred husk of your family's home, you can trade in your life savings for passage into the galactisphere.

A benefit of this arrangement is that if you're paying, you're a customer. In a way, that means the captain and his crew now work for you. Just how far your currency will get you remains to be seen, but if there's an important disagreement, don't be afraid to use your meager leverage. The crew may even see that you're better equipped to be their leader and stage a mutiny—which has the added benefit of propelling you instantly to Space captain status. From there, you only need a few quick heroic deeds and death-defying situations to make Space Hero.

No credits? Consider indentured servitude. Do you know

any ballads? How is your mop grip? Dig deep. Willingness to endure emotional scarring is its own form of currency.

⊕ Buying Passage Pros:

- Travel in relative luxury
- Rightful access to ship's food and water
- No fear of unexpected Space walks
- Potential to make lifelong Space adventure friends
- Complimentary toilet access

⊖ Buying Passage Khans:

- Buying is rarely as satisfying as stealing from no-goodniks and Space hippies
- Appearing on official manifests makes you a bigger target when ships are boarded by authorities, slavers, and Space pirates
- Spending all of your life savings or ill-gotten gambling gains means you'll have little left for room, board, or non-gag-inducing nourishment when you arrive at your destination

5. Wait for Your Planet or Colony to Come under Attack, and Then Survive

Space is dangerous, but so are far-flung planets. Because you know what surrounds planets, don't you? That's right—Space. And it's full of awful aliens, marauders, slavers, and radioactive waves that prey on the weak and innocent.

But there is an upside to everyone you know becoming

baby alien incubators or sex slaves for various betentacled creatures. Since you're looking to become a Space Hero and must
therefore be exceptional, there's a good chance you'll survive
your encounter with such colony-massacring horrors. That's
good news! Take comfort in the knowledge that when a
colony is steamrolled by some extraterrestrial armada, the local
Space government usually dispatches help. Get found by that
help, and you'll have an instant ticket off your home world for
the low, low price of everyone you've ever known or loved.

⊕ ATTACK SURVIVAL PROS:

- ◉ You survived!
- ◉ FREE ride off the podunk planet you once called home
- ◉ FREE nemesis (or nemeses)
- ◉ FREE tragic backstory

⊖ ATTACK SURVIVAL KHANS:

- ◉ Everyone you know is dead
- ◉ High probability of (FREE) physical injury
- ◉ You can never go home again
- ◉ Potential to fail the "surviving" portion
- ◉ Really not much of a foolproof method for achieving
  Space. In fact, pretty much the worst method. Even worse
  than stowing away.

6. BE BORN IN SPACE LIKE SOME KIND OF ALIEN FREAK

Occasionally, children are born aboard starships and merchant
vessels. If this is you, congratulations! You've sidestepped the

hurdle of achieving escape velocity from your own boring terrestrial life.

Unfortunately, you've still got that whole exceptionalism portion to master. After all, you were born in Space—you didn't *earn* Space. That means your chances of being a Space sanitation officer or a Space disc jockey are much higher than your chances of being a Space savior of the innocent or a Space smasher of hostile alien faces.

Apart from the other potential spacefaring Heroes making fun of you, a Space birth has its own issues—of the developmental deformity variety. Since humanity (as well as just about all other life) evolved in gravity, it turns out that gravity is an essential part of natural development. Skeletal structures, primarily, require gravity to help with alignment and bone density. But muscle weakness and even brain deficiencies are known to crop up in Space babies born without gravity's helping, downward-pulling hand. Ipso facto, if you want to be a Space Hero, you're going to want to get at least some simulated gravity under your ass.

If you *were* born a Space love baby, all is not lost. The basic tenets of becoming a Space Hero still apply, even if you've skipped the messy business of finding your way out of your pitiful gravity well. You'll need tragedy, gumption, a mentor, and all the rest of it, with the added caveat that you'll need to be even more interesting, since for you, Space is humdrum. Time to get into a bar fight with some Trandoshans or invest in cybernetic legs or something. I can't do all the figuring for you—be creative!

 ## SPACE-BORN PROS:

- You're already in Space!

- The rigors of Space survival are your way of life
- You can steal that ship you're on

⊖ SPACE-BORN KHANS:

- You didn't *earn* Space
- Boring childhood
- Lack of affection for ground beneath your feet and a sky over your head
- Potential for lack of muscle mass and bone density and for brain damage if in zero-gravity environment
- People will think you're an alien-like freak

7. BLAST YOUR WAY THROUGH THE ATMOSPHERE!

You're an enterprising lad or lass, unwilling to be held back by the constraints of your circumstances! So there's no intergalactic military to join, no pirate crew to which you can pledge yourself, and no airtight cargo container to transform into your own cramped waterless water closet—so what? Create your own way off the boring rock that begot your existence!

Thankfully, Earth's twentieth century was filled with government agencies from various countries that pioneered the nature of doing things in Space—like pushing important buttons, giving others orders, and voiding bowels in microgravity—so that you don't have to! Stand on the shoulders of the folks who figured out how to get toilets to work without gravity, and get to it, Space adventurer! You have a rocket to build.

THE SPACE HERO'S WORKBOOK:

What Is Your Heroic Tragic Backstory?

_____

What Is Your Mentor's Name?

_____

How Did You Get into Space?

_____

EPISODE II

THE FINAL FRONTIER

In the beginning, the Universe was created. This has made a lot of people very angry and has been widely regarded as a bad move.

—Douglas Adams

CHAPTER 3

SPACE: A PRIMER

Say, you're in Space! Nifty. Here are some things you should know:

IT'S BIG

You'll want to begin your move toward Space fame with baby steps. The Universe is still largely unexplored, but the portion observable from Earth is close to 92 billion light years in diameter, or approximately $4.9007241711711711711711711117 \times 10^{26}$ dead ensigns laid end to end. Many courageous men and women (and some not-so-courageous convicts humanity eventually sent to colonize planets against their will) have vanished into the icy depths of the Universe's bosom. If you're not careful, you are likely to join their ranks: another forgotten fleck of Space dust, hurtling through nothingness for all time.

On the starry side, even a "baby step" relative to the Universe's great breadth is more than you could possibly hope to explore before something goes fatally wrong. If the Dead Ensign Scale isn't doing it for you, consider this: all of the known Universe and all of the speculated Universe is less than 5 percent of the Universe's mass. The rest is "dark matter," which we only know enough about to claim ownership over and, in some

cases, exploit for fuel. Otherwise, it's rather fearsome and neb-
ulous. What's in the dark matter? Parallel universes? Matter that
can only be perceived by more ancient, more godlike creatures?
Seedy buffets? If I had to guess: yes.

IT'S COLD—UNLESS IT'S HOT

Atmospheres and planets are warmed by stars. The matter that
makes them up absorbs stars' radiation. In Space, the void-like
nature of the Void means there's nothing between you and solar
radiation and also nothing else to absorb that radiation. The result?
The Void is really cold or else really hot, depending on where
you are.

In the direct light of a relatively nearby star (say, when you're
in orbit around Earth), you can encounter temperatures of 120
degrees Celsius, or 248 degrees Fahrenheit. Meanwhile, on the
dark side of a planet or moon, it can get colder than the under-
side of a billion pillows—as low as −100 degrees Celsius, or
−148 degrees Fahrenheit.

As a point of comparison, hypothermia sets in when your
core temperature is lowered to below 90 degrees. *Hyper*thermia
(the too-hot inverse of hypothermia) becomes an issue when
your core temp is up to about 100 degrees. In short, your Space
suit has to shield you from those extremes. For more on Space-
suit particulars and parameters, see Chapter 6: Clothes Make
the Spaceman.

IT'S A VACUUM

Though there have been a number of instances in which humans
have survived short-term exposure to Space's icy vastness (some
of the more famous instances being those of Senior Chief Petty
Officer Galen Tyrol and Petty Officer Second Class Cally Tyrol of

the Battlestar *Galactica*, astronaut John Crichton, Earthling civilian Arthur Dent, and Betelgeusian Ford Prefect), it's never pleasant. And that's not just because it can be cold (or hot) or because you'll pass out from a lack of oxygen and asphyxiate in a matter of seconds. You'll also have to contend with the boiling fluids of your own body.

Atmospheres are made up of air, and that air presses against you all the time, exerting force "inward" on your body. To counteract that force, your body has a natural level of pressure that pushes "outward." The result is an unnoticeable balance, more often than not. This is why airplane cabins are pressurized and why your ears pop coming down from a large hill back on Earth, where the pressure is greater at the bottom of the hill than the top. Your body adjusts.

But a lack of atmosphere in Space means no pressure in Space. A sudden change from pressurized vessel to empty vacuum is too great for your body to make its usual ear-popping adjustment. Instead, you get a mess of problems. The boiling point of most liquids drops to less than your body temperature while in a vacuum, which means your eyes will boil, and your tissues will expand as the rest of your fluids boil along with them.

The good news is that, like asphyxiation in a vacuum, it takes some time for this to all happen—nearly a full minute! It's possible to survive very brief exposures to the Void, as long as you get back to oxygen and pressure as quickly as possible.

IT'S IRRADIATED

If the incomparable, ceaseless, soul-crushing, ex-spouse-like temperature fluctuations weren't enough, you—being a future Space Hero and therefore attentive to detail—may have noticed that Space is without protective planetary atmospheres, magnetic fields, and ozone layers. This can only mean one thing: Radiation. The silent killer.[1]

Your and your crew's attention will be occupied by actively attempting to avoid losing yourselves in the unchanging landscape of glittering blackness, succumbing to Space madness, or freezing to death because some idiot decided to test the limits of your life-support systems. During all that time, Space will be passively trying to kill you by blasting your cells into submission.

Galactic cosmic rays,[2] or GCRs, are proton-heavy atomic nuclei accelerated into Space by suns, stars, and solar flares.

With no atmosphere or magnetosphere to protect you, the invisible particles travel uninhibited through the cosmos, through your ship, and even through your cell membranes. Even if the effects are not acute, cosmic rays can permanently damage your DNA and can cause regular cell and brain cell death. They can even cause irrepressible vomiting.

Primitive astronauts were exposed to as much as 500 to 1,000 millisieverts of cosmic radiation in a yearlong Space journey, even within the confines of their vessels. That's

---

1. Except for all the vomiting, which is pretty loud.

2. Not to be confused with Cosmic Ray's, a great strip cantina in Sector 12.

compared to the 2.4 mSv the average human would have endured on Earth during the same period. At around two hundred to four hundred times the normal barrage of GCRs over long periods of time, crews would find it considerably more difficult to "poo poo in the potty," much less make complex calculations and interpret flummoxing intergalactic navigational charts. So you see, radiation is all around you, and radiation protection should be too.

ANTI-RADIATING YOURSELF
DETECTION PROTECTION

To avoid getting your DNA violated by Space's dirty, lecherous high-energy particles, listen to the sultry voice of your ship's artificial intelligence construct when it warns of unsafe levels of exposure. Your ship's AI knows this because your ship's AI knows everything.[3] Just ask Dr. Dave Bowman, who succumbed to Space madness after defying his then state-of-the-art HAL 9000. If your ship is not equipped with an AI, the crew's doctor will have to monitor all crew members' radiation exposure on the regular. If your ship doesn't have a doctor, the crew member with the most field-medic experience or the fewest responsibilities should be appointed interim doctor by the captain.

HIDE FROM THE RADIATION

You've detected an incoming solar flare or GCR wave. You have precious little time to prepare. Unfortunately, just like Cosmic Ray the barkeep, galactic cosmic rays aren't easily placated by walls, running, or promises to pay that gambling debt in just a few more days' time. Even with modern materials, deflecting cosmic radiation is difficult, as GCRs have a tendency to hit the

---

3. Remember: Everything you say or do during private bunk or toilet time can and will be held against you at mandatory crew holiday parties.

molecules that make up radiation shelters and kick off even more deadly particles.

It's kind of like hiding from the effects of a giant microwave inside a smaller, more powerful microwave. Without, that is, a glass window to help catch your splattered remains. At least with a solar-flare warning system, you can buy yourself and your crew some time to get the hell out of the way of the rays and avoid the radiation's most concentrated doses.

SHIELDS UP!

Atmospheres of planets dispel dangerous radiation due mostly to their densities. High-energy particles from the sun bounce off air molecules surrounding Earth, for example, and you get a tan instead of a searing radiation burn. Water is a high-density material that's quite effective at warding off radiation. If you can manage to sustain a security blanket of water around your ship at least a few feet thick (perhaps as a layer of the hull), the H_2O molecules will slap that radiation around like a battleship full of drugged-up starfighters bracing in sleepless anxiety for the next wave of attacks by the relentless, emotionless enemy. For this reason, along with several others better left to the imagination, a water bed is recommended as the centerpiece of captain's quarters decor.

Trouble is, a water shield greatly increases the mass of your ship and makes propelling it even more difficult. There are other, less effective materials, like layered polymers with radiation-absorbing coatings that can be both lightweight and finished in black cherry or bitchin' flame paint jobs. It's best, however, to mimic a planet's electromagnetic field with an energy shield. It'll deflect high-energy particles and GCRs and hopefully the occasional laser beam as well. As they say in the former French Republic of Crater: *Ta da!*

TAKE YOUR MEDICINE

Advanced alien cultures have determined herbal as well as phar-
maceutical means of living with regular doses of radiation. Some
even find that being slightly irradiated and occasionally glowy can
be quite beneficial. But unless you're an alien, you're not an alien.
Your safest bet lies with avoidance of radiation. If you have to,
taking plenty of anti-radiation medication can help, but it's not
always an effective solution. Potassium iodide, for example, can
offer protection against the radioactive isotope iodine-131—and
comes with complementary nausea and barf bags—but its effectiveness is pretty limited. Thanks to aliens, leaps in medical technology offer better alternatives. So remember: always say yes to
Space drugs.

CHAPTER 4

Unless you've commandeered one, it's unlikely you're captaining or traveling on a five-star interstellar cruise liner. If you have any hope for survival, a few baseline resources are required to safely and effectively maintain your existence.

FUEL

The bad news is that fuel is heavy, expensive, and difficult to come by. The other bad news is that, with no friction in Space, you'll need fuel to accomplish a lot of basic maneuvers. Every change in direction or orientation requires you to not only burn thrusters in order to move your ship, but also to *stop* moving your ship. On any given voyage, the first half of your trip is spent accelerating toward your destination; the second half is spent burning fuel to bring your ship to a stop.

But one of the biggest expenditures of fuel will be used in breaking a planetary atmosphere. Gravity, though considered a "weak force," holds a powerful and relentless grasp on your comparatively puny rocket. To break away from your planet's clingy grasp, you have to reach a minimum escape velocity, which depends largely on the planet's size.

"Escape velocity," however, is a misnomer—one of the few

times Science gets it wrong. What you're actually achieving is an escape *speed*: a speed at which the combined kinetic and gravitational potential energies of an object (your ship) reach zero. Achieving such a speed while taking into account the weight of fuel would seem to create a physics-borne paradox. The greater the gravity of the planet, the greater the escape speed required, and therefore the more fuel required—but the more fuel you bring, the heavier your ship and the more difficult it is to achieve escape velocity. Congratulations: you're stuck on that planet forever.

Or would have been, if not for the Oberth effect.

The Oberth effect describes the phenomenon of a rocket engine (and the rocket fuel therein) generating more useful energy at higher speeds than at lower ones. Put another way, as the speed of your ship increases and the effect of gravity decreases (you are gradually leaving its influence the farther you travel away from the planet's surface), less fuel becomes capable of greater work. You won't be stuck on Narn forever after all!

That said, whenever you land on a planet, you need to make sure you're able to get off again. This is why most captains keep their ships comfortably in the planet's orbit and send smaller (lighter) recon ships down to planet surfaces. If you've got friends in accounting, consider the installation of "up-beaming" teleportation technology. You can invoice the SOS at a later date.

For more on Space fuels and engines (nuclear power sources; FTLs, or faster-than-light drives; warp drives; and the like), please refer to Chapter 11: Warp Drives and Space Fuels.

SPACE FOOD

There is not a creature alive today who remembers the ancient subspecies of humans known as "Vegans." The Vegans lived by a strict dietary code that barred them from eating food made of or from any animal. There is some debate in the history holodrives about how the Vegans died off. Was it because the proliferation of Space travel made such dietary pickiness unsustainable or more directly the result of the 2084 discovery that plants are sentient and in fact feel a great deal of pain when steamed?

Whatever the reason for the Vegans' eventual extinction, the holodrives agree that the Vegans were oppressed throughout their existence for two primary reasons:

1. They spent a great deal of time attempting to re-create non-animal-based equivalents of the animal-based foods they had vowed not to consume.
2. Animals are delicious.

The sad reality, and one for which humans and aliens will never be able to apologize, is that with their attempted creation of real food facsimiles, the Vegans were actually on to something. Willingly avoiding delicious animals, however, still defies the comprehension of this red-blooded meat eater.

Studies have shown that the greatest benefit of food in Space isn't nutrition or even taste. It's *comfort*. The sweet aromas flowing from the family kitchen. The thrill of cheating on a diet with a slice of peach cobbler pie. Licking peanut butter off a Spacehooker's tentacle.

Nutrients are easy to replace, but there's no substitute for the flavor of memory. Comfort food's importance to crew morale cannot be overstated. So when considering your stock of Space pills, Space slurries, and freeze-dried Space food, don't be afraid to splurge a bit on a supply of your crew's favorite home-cooked meal or a case of facon (a plant-based recreation of an Earth delicacy derived from the now extinct pig,[1] known as "bacon"). It could buy you your crew's love and blind devotion. If you're not careful, they may even confide in you.

However, food has hidden dangers—specifically when it goes flying through the confined inner space of your ship.

IN THE EVENT OF AN ARTIFICIAL GRAVITY MALFUNCTION...

You are all manner of screwed. Even the smallest floating bread crumb can become lodged in your ship's computer core or

---

1. Which, coincidentally, died off at nearly the same rate and time as the Vegans.

air-filtration system and unleash havoc. That's to say nothing of condiments, libations, and half-chewed beef particles that have escaped because you never learned to chew with your mouth closed.

Some precautions to take until the gravity drive is up and running again:

1. No Snacking

All eating must be monitored, regulated, and contained to the mess. Mess doors should remain closed while eating is in progress.

2. No Crumbs

A ship-wide pill diet should be adopted. Once the pills are gone, draw straws. The drawer of the shortest straw is to be eaten…*after* you make your way through the rest of the food rations. Start with the most solid foods (freeze-dried ice cream, bite-sized ration bars, real and fake meats) and work your way down to the most dreaded of Zero G grub: the corned beef sandwich.[2]

3. No Carbonated Beverages

Without gravity's aid, the human stomach can't separate liquid from gas in the stomach. The result is a form of bile-surgy wet burp that would make even the grossest baby embarrassed on your behalf.

OXYGEN

That's right: You need to be able to breathe in Space. And though your Space suit serves as a suitable backup for a contained environment in the (fairly likely) event of a hull breach, in general,

---

2. One such corned beef sandwich, smuggled aboard the *Gemini III* in 1966 by astronaut John Young, nearly brought down Earth's entire Space program almost before it began.

the insides of your Space vessel should be filled with breathable air. For those among you and your crew who can be considered human, humanoid, or Earthian in origin, this means oxygen. Which does *not* mean pumping the whole pressurized cabin full of the stuff. For one, it'll make such a pressurized cabin more flammable and grenade-like. Also, too much oxygen eventually causes oxygen toxicity, a medical condition that can lead to a vast array of problems.[3]

The makeup of Earth's atmosphere—and therefore the ideal composition aboard your ship—is approximately 78 percent nitrogen and only 21 percent oxygen. The final percentage point should be made up of water vapor, which makes breathing air a bit less like drinking sand.

Despite the dangers of too much oxygen, not enough oxygen is worse, which is why your ship needs to be constantly producing it in some form or another. Ideally, you'll be looking for a closed system that can replace what you breathe out with what you need to breathe in. On planets, plant life that performs photosynthesis and expels oxygen makes up the other half of the equation when paired with animals that breathe oxygen. Bringing along plants isn't always a viable option in Space, however. The amount of plant mass needed to keep even a small ship and crew in oxygen indefinitely is huge.

Other methods of getting oxygen in your ship include water electrolysis, which splits water molecules into oxygen and hydrogen molecules. In early Earth Space exploration, the extra hydrogen was vented into Space. Later, it was combined with carbon dioxide to create more water (along with methane, which was vented) that could then be split again.

---

3. It's a whole helluva lot harder to stop the Klingon boarding party currently eviscerating you when the air has detached your retinas.

Of course, your crew will be breathing *out* as well as *in*, which means the balance of your environment will gradually turn from oxygen to carbon dioxide, the other silent killer.

CARBON DIOXIDE AND OTHER TRACE GASES

Breathing is a dangerous dance. Every time you breathe in, you suck up the oxygen you might need later, and what you blow out is carbon dioxide, which is essentially a gorram poison. The more carbon dioxide you're breathing in, the less oxygen you're breathing in, and the less oxygen that is available. That's to say nothing of the other trace gases your body produces that are *actually* poison—like methane, acetone, ammonia, and methanol. All those have to be scrubbed out of the air before they build up in your ship and murder you.

Carbon dioxide poisoning, or hypercapnia, sucks and burns and flays the ventricles of your being. As the gas collects in your spacecraft, it can build up in the blood. Early symptoms of carbon dioxide poisoning include headaches and lethargy; later ones move on to disorientation, convulsions, panic, and unconsciousness. In effect, carbon dioxide can trigger bouts of a form of Space madness, and the only thing worse than suffocating in a tiny metal cylinder is getting shot to death by a psychotic crewmate before *he* suffocates in a tiny metal cylinder.

We've no shortage of documented incidents in which a crew found itself losing its mind in the Void, but some of the most notable are also the most horrific. Consider the fate of the *Event Horizon*,[4] which reappeared years after vanishing through a black

---

4. While rumors circulating following the rediscovery of the missing *Event Horizon* suggested that it had somehow traveled to hell, which had driven its crew insane, a mixture of carbon dioxide poisoning and bad captaining ultimately led to the crew's demise. Hell indeed.

hole, filled with the mutilated corpses of its crew.[5] The *Icarus I* mission to jump-start Sol (the cool kids' name for Earth's sun) after some trouble with nuclear reactions met with a crazy-murder fate[6] as well, proving somewhat empirically that blindness isn't the only reason you shouldn't stare into the sun.

The point being you'll need not only a way to maintain oxygen levels in your ship—and that means additional air in storage or some means of manufacturing O_2—but also a way to get the carbon out. If you're going to have a seizure during cease-fire negotiations with the Romulan government, it had better be a clever ploy to buy more time or create a diversion, not because of a dirty air filter.

"STAY STUNNING! WITH DIRK PARSEC

Getting the Most Out of Your Crew

Is your crew sluggish? Terrified? Depressed? Pushing seventy straight waking hours as the enemy descends upon you every thirty-three minutes like clockwork? Though maintenance of the 78-21-1 nitrogen-to-oxygen-to-water vapor ratio is generally advisable, I can personally attest to the benefits of occasionally "tampering" with the ship's atmospheric controls.

If we learned one thing from Las Vegas and its lawless follow-up, Planet Vegas, it's to not fall asleep

---

5. Which is stunningly unfortunate when you consider how difficult it is to survive passage through a black hole in the first place.

6. Here's the thing about Space travel: it's like *The Shining* up there. You're cooped up with people you grow to resent and hate, and you constantly risk flying into a murderous rage. (See also Chapter 9: Sidekicks and Ragtag Crews.)

with a hooker in your room unless you want to awaken credit-less and without a change of pants. But if we learned two things, the second is that pumping near-dangerous levels of oxygen into the air makes people happier, luckier, and less dependent on sleep, mankind's sole innate weakness. Sometimes the ends of increased productivity justify the means of potentially poisoning your crew. So while you shouldn't keep your crew hopped up on pure oxygen *all* the time, that decision (which you should not share with them under any circumstances) is yours.* After all, you're the captain.

*Note: There's no hard scientific evidence that increased oxygen will make your crew more efficient. Also, casinos pumping oxygen into the air supply to fuel additional gambling is an urban legend. It certainly had nothing to do with the Planet Vegas planet-wide explosion of 2601. —Ed.

FECAL MATTER: SPACE'S WONDER SUBSTANCE

Everything on your ship takes up room, adding to the ship's size and mass, and becoming a potential liability. To make every bit of mass aboard worth its volume and weight, it all needs to pull double duty. It needs to perform not only its primary function, but also a secondary function, like help in the ship's fight against pressure issues, escape velocity issues, life support issues, and more.

Everything you bring with you into Space is important. Even things you didn't realize you brought: namely, your poop.

Here's the thing about the waste products your body produces: they actually include a lot of things your body could still use. Urine, for example, is mostly water, which is hard to come by in Space. That's right: you're going to be drinking (filtered) pee, pee-drinker.

As for fecal matter, it's harder to regain any lost nutrients from poop, but all that solid waste has to go somewhere. Sure, you could just jettison it out into Space, because Space is huge, like the ocean, and it's unlikely anything or anyone will ever come into contact with it.[7]

An even better use, though, is as a giant poop shield.

As you'll recall, or have found out the hard way, Space is filled with harmful radiation. That's where your poop comes in after your poop comes out.

Radiation is energy, and energy travels in waves. Metals such as lead do an okay job of stopping radiation, but aren't practical additions to your spacecraft's hull. That's because it's vital to do all you can to keep mass to a minimum. An even better radiation shield is made up of the kinds of molecules that are dense, but moving, such as those found in water. A wave of radiation has a greater chance of being bounced back into Space by water molecules because they're abundant and moving somewhat rapidly.

Food and poop can work in the much the same way. Neither substance holds the radiation that hits it, unlike your wussy cells. Instead, the radiation bounces off. You can pack your radiation shields with food, then eat that food and suffer no ill effects from eating it. A more permanent solution is to replace your food slowly with the fecal matter you create from eating it. Why throw all that shit away when it could be saving your life?

And thus, you've just learned an extremely important lesson

---

7. Although if something *does* come into contact with it, it won't be the "poop on Space windshield" laugh riot you're imagining—it'll be a catastrophic impact from a deadly, high-speed frozen poop missile.

about Space: one man's crap is that same man's treasure. Where the average Spacefarer might see disgusting sewage, a real Space Hero sees a solution to a seemingly unrelated problem, a wave of radiation repelled, and another day of awesome Space adventures ensured. So don't waste waste. It's a waste.

ARTIFICIAL GRAVITY

Gravity is, believe it or not, a vital ingredient in your biological well-being. Just a few days' time in Zero G and your bone density will begin to drop. Your entire body begins to weaken. Space Heroics *require* dense bones. And so, per standard operating procedure, what we don't have in Space, we fake in Space.

Your ideal relationship with gravity is like the one between you and a crazy aunt. You only have to think about it when there's a problem. On most ships and Space stations, the transition from planetary grav to artificial grav is seamless. But in case it's not, and you have to go pick crazy Aunt Grav up from the promenade for streaking through the food court again (metaphorically), it's best to know what's keeping your designer Space boots on the floor beneath you so you can fix it.

CENTRIPETAL GRAVITY

Some ships and many orbital Space stations create artificial gravity by rotating continuously. Their circular or cylindrical designs mean the floor of the ship or station is actually the curved inside of the outermost hull. The "outward" force keeping your feet on the floor—referred to as centri*fugal* force among Space idiots—is actually an illusion.

What you're really feeling is centri*petal* force—your inertia, tangential to the direction of the continuous spin of the ship—and the floor pushing back beneath you, stopping you and everything aboard from flinging off into Space. This is Newton's third law of motion at work, which we'll get to in Chapter 18, in which Isaac Newton, known as Long-Nosed Ike in Space boxing circles, rustles your jimmies with some seventeenth-century Zero G fighting prowess.

Centripetal gravity, though once common, was eventually abandoned because of its messy side effects. For starters, it's not really gravity, but the illusion of gravity. It doesn't even work in the same direction. Real gravity pulls mass inward, but centripetal gravity sends it out. This, combined with the much smaller size of the stations as compared to planets and other gravitational bodies, means that the "gravity" felt at your feet is much different than what's felt at your head.

It makes maneuvering through the environment an awkward adjustment, which defeats half the point of having artificial gravity in the first place. Jumping and falling, for example, can cause the Coriolis effect, a disorienting nausea (not unlike that caused by kissing a crazy aunt) from the feeling of being pushed outward. Welcome to life inside the merry-go-round from hell, yet another puke-inducing mainstay of Space.

GRAVITY PLATING

Through advancements in Science, artificial gravity in most ships is now installed in the floor plating. Gravity generators produce gravitons—the subatomic particles responsible for gravity—without the need for corresponding mass. The power of the gravity generators can be adjusted by plate, room, deck, or ship. Perhaps you fancy some high-grav training in the gym, for squats that produce double dividends, or would like to implement low-grav casual Fridays.

The only real side effect of gravity plating is that the

STRANGLING AN ALIEN

SOPPING UP LEAKS

AS A BLANKET

SIGNALING A SHIP

AS AN ACTUAL TOWEL

PLUGGING A
HULL BREACH

AS A DISGUISE

KEEPING THE
CREW IN LINE

technology is related to—and can therefore sometimes interact with—your ship's warp drive technology, should you have some. For solutions to these problems, and for recommended workout regiments in high or low gravity, check out *Synthetic Grav Field Integration Considerations* by Dr. Leah Brahms.

TOWELS AND THEIR APPLICATIONS

By now you've grown to know the importance of a towel. As was detailed extensively by the work of famed futurist and soothsayer Douglas Adams, the towel can be an incredibly useful object in Space. Apart from helping you to soak up excess moisture, towels can serve as makeshift clothing, can help hide your identity, and can be used as weapons if you find yourself in an "alien prison break" scenario.[8]

Towels, as we all know, also convey a useful psychological boost, as most Spacefarers assume that anyone who manages to hold on to a towel probably has other provisions, including a blaster pistol, a backup blaster pistol, and a good number of allies. Just about any Space Hero who can manage to hang on to his or her towel is assumed to be a Space Hero to be reckoned with.[9]

---

8. "Wet-rolled and whip-cracked," so goes the motto of the Prison Break Pirates.

9. Always try to be the kind of frood who really knows where his or her towel is.

EPISODE III

THE UNDISCOVERED CAPTAIN

I don't believe in the no-win scenario.

—James T. Kirk

CHAPTER 5

WHAT SPACE CAREER IS RIGHT FOR YOU?

S pace is about efficiency. Running a tight ship. *Compartmentalization*. The few spaceships that make it beyond their maiden voyages without becoming either Space junk or Space coffins compartmentalize *everything*. Compartments for water, compartments for food, compartments for sleeping. There are even compartments for the planning of other compartments. This stringent level of organization begins with the crew. All of the compartmentalizing in the 'Verse is for naught unless every crew member understands his, her, or its role: the compartment of their duty. Which is not to be confused with the compartment of their dooty, which the lowest-ranking crewman should remember to discharge at regular intervals.[1]

Lucky for you, finding the compartment of *your* duty is easy: it's whatever compartment is the best one, with the feather pillows and the wet bar if you've procured a ship worthy of your Space Hero aspirations. But there are many kinds of heroic compartments. What kind of ship are you running? What kind of life would you choose for you and your eventual crew? How shall

---

1. Unless it's your duty to use dooty for other things—in Chapter 4: Space Resources and Other Space Shit, see "Fecal Matter: Space's Wonder Substance."

you compartmentalize *yourself*? This is a more practical decision than the fabric-of-your-being existential personality quiz posed to you in Chapter 1. Below are the best Space careers for you and your compartment-filled ship.

MAVERICK MILITARY COMMANDER

The best way to see Space and also have access to a large battery of incredibly powerful weapons is still to rise through the ranks of your local Space military. Why bother with buying, building, or doing when the government can buy it, build it, and do it for you?

Of course, this may not be your style. After all, militaries have rules, chains of command, politics, and rules again. Space Heroes are not known for their ability to follow rules. In fact, they're usually known for their complete lack of adherence to authority.

Also, depending on your governmental allegiance of choice, uniforms may present an impediment to exuding your amazing sense of fashion. (For more on Space Hero fashion, see Chapter 6: Clothes Make the Spaceman.)

However, the military can offer you many upsides, especially if you're willing to put in the career time. You get sanctioned fighting to help you cut your teeth in Space battles and hone your command skills, provided you survive. You get the choicest cuts of meat and the best bootleg Space hooch. And at the end of it all, you get a big-ass ship filled with minions who are literally paid (by someone else or taxpayers or both) to listen to you.

And as a maverick military commander, the opportunities for Space Heroism are considerable. Look at the pinnacle example: James T. Kirk. Kirk was a well-loved captain who had it all. As commander of the USS *Enterprise*, he was able to lay waste to

WHAT SPACE CAREER IS RIGHT FOR YOU? 57

enemy ships, beam down on whatever planet he liked, and bang all sorts of Space babes, all while losing members of his crew on *nearly every mission*. The guy's crew survival rate was abysmal—and yet he's loved by literally billions.[2]

"STAY STUNNING!" WITH DIRK PARSEC

The Best Quarters on the Command Ship

Within that big-ass ship, the most badass quarters are those of the captain. Decorated to your liking, the captain's quarters can be an essential sanctuary from the rigors of leadership and for the rigors of sexual conquest. USS *Enterprise* Captain Jean-Luc Picard decorated his with tasteful plants and 1980s-reminiscent furniture designs,* Commander William Adama on the *Galactica* had a model clipper ship,† and Commander Shepard's SSV *Normandy* quarters included a giant glass dangerous-to-have-in-your-warship's-most-important-bedroom aquarium.‡ It's fair to say that your options for an amazing bedroom (and amazing potential activities to take place in that bedroom, *wink*[B]) are unlimited as a maverick military commander.

*Captain Picard's ready room also included a number of awesome model Spaceships, which he surely used to stage mock Space battles while he sat there by himself waiting for stuff to happen.

†A ship commander Adama subsequently destroyed during a particularly dramatic and heartfelt emotional outburst, much to the horror of the last remaining historians of the human race.

2. That's not even counting the illegitimate green children he has scattered all over Space.

‡Generally, spaceship commanders don't encourage keeping objects that could explode into shards of deadly glass shrapnel and fish poop, but Commander Shepard was a consummate rebel.

ß *Dirk's Tip:* "Want to see my aquarium?" gets 'em every time.

STARFIGHTER JOCK

One of the more prolific jobs among Space folk is that of pilot. Most everyone should at least be moderately versed in how to pilot a spaceship, given that you'll spend your entire life aboard one. And if you're going to interact with other Space folk while in Space, chances are you're going to do so angrily and with lasers. Enter the starfighter: a small, quick ship best equipped for laying down sparkly death. (You'll learn more about the starfighter and other spacecraft in Chapter 8.)

The starfighter jock is a Space Hero who's not afraid to get his or her hands dirty. Even if your glory days are behind you, as a starfighter jock, you never pass up the opportunity to grab the helm during a death-defying escape or hop into a fighter and see to it personally that the starry skies are painted with the blood of your enemies. The biggest victories in Space battles are rarely won by the commanders of capital ships.[3] Instead, the victors are the farm kids from backwater desert planets and the emotional cripples who happen to be pilot savants, duking it out in the figurative and literal trenches of Space stations.

Your innate reckless abandon means a small crew is preferable, and you require a crew that's competent enough to manage itself while you're off deck. What you lose in hands-on leadership, you make up in inspiration and setting an unattainable

---

3. Those guys are usually too busy yelling obvious shit, such as "It's a trap!" and "Concentrate all fire on the ship that's brutally kicking our asses now that we've fallen for this rather obvious ruse!"

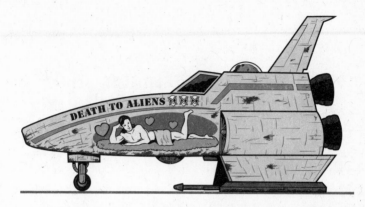

example. If you find yourself butting heads with the young hotshot starfighter jock (or you *are* the hotshot starfighter jock looking to usurp your stick-in-the-mud commanding officer), you're on your way. If you are consistently the only one who believes you'll make it back alive in the face of imminent death by explosion—in the nick of time, no less—then arrived already, you have.

Making a business out of the starfighter jock lifestyle is maximized by playing loose with alliances. Is there an empire to take down? A blockade to be broken? A rogue alien species that deserves extinction? As long as there are ships to be shot down, there's work for you and your crew. Follow whatever vague moral code helps you sleep at night, but recognize that if you want to keep that ration compartment full, you can't be above the occasional mercenary gig.

If there's a downside to the starfighter jock life, it's the horrific mental toll it takes on most of its most shining examples. Starbuck's Curse, as it is known, sees to it that every pilot afflicted with the venerated call signs of "Ace" or "Hot Shot" eventually devolves into an unmitigated, impossible-to-love, cigar-addicted jerk face battling bouts of substance abuse and post-traumatic stress disorder[4] in an unending cycle.[5]

TOUGH-AS-NAILS FREELANCER

The frequency with which people in Space want other people in Space dead may alarm you. Who's going to stop them? The Space police? Ha! But Space is big; that guy they want dead is far, far away; and Hutts don't often fit into ships without large cargo-bay doors and loading forks. Yes, to see this particular breed of scum dead, those Hutts need to turn to an expert. They're going to need a freelancer. Some may spit on the sand you walk on and call you a "bounty hunter," but if they do, you should kill them. Besides, you're a good guy.

And though offing folks for credits takes its fair share of groinal fortitude and a decent hit percentage with a blaster, the reality of the bounty hunter's life is that killing is the easiest part of the job. The "hunting" part of the job title is deliberately misleading. Most of what you'll be doing and need to be good at is…traveling.

You'll have to travel to get a mark, travel to find that frakker, outtravel other bounty hunters to find that frakker first, pursue that frakker when he or she runs, travel to threaten that frakker's loved ones (if applicable), and travel back to the client with the body of that frakker (remember: no disintegrations) to collect your

---

4. PTSD makes it hard to realize that the attack ships you're seeing on fire off the shoulder of Orion actually have your friends and comrades inside, and that they are both burning *and* suffocating at the same time.

5. The famous leader of the Rebel Alliance Rogue Squadron, Wedge Antilles, was known to be incredibly cool and personable in the cockpit— but once out of it, rumor has it he would disappear on week-long sex binges with wookers (wookiee hookers). Other rumors have it that he just liked to talk and cry the whole time.

bounty. Your crew will need to be small, loyal, and reliable in a pinch—but also entertaining and at home in the boring blackness of infinity. Not *too* entertaining, though. No one likes a self-appointed comedian.

If you choose to go it alone—which, let's face it, makes you somewhat less vulnerable and more darkly mysterious—stock up on books, movies, and fuel. It gets awful lonely out there in the black, especially if you forget to bring porn.

On the plus side, by necessity, bounty hunters often have access to the coolest gadgets. After all, what's scarier—a guy or a *guy with a metal jet pack flying through the air coming at your face*? (Although you're probably better off looking like an unsuspecting regular Joe so people don't suspect you, run from you, or shoot you on sight. But keep the sweet jet pack for personal use on your downtime.)

"STAY STUNNING! WITH DIRK PARSEC

The Pros and Cons of Jet Packing

During a brief stint as a bounty hunter scouring certain distant planets, as well as through other heroic Space adventures, I've had a few opportunities to use the storied and coveted jet pack. I'm of two minds on its value as a bounty hunter's tool. On the one hand, it's bad luck. Infamous father-and-son bounty hunters Boba and Jango Fett both fell victim to jet-pack failure at critical moments. Boba Fett found himself in the belly of the Sarlacc, thanks to a fairly routine, expected jet-pack malfunction that exposed a massive design flaw in the supposed lifesaving device.

A jet pack also failed Boba's dad, Jango Fett,

who was unable to use his jet pack to evade a guy walking briskly toward him with a sword.

In fact, with all the personal flying vehicles around, a jet pack feels rather redundant. It's also among the least efficient in zero gravity, since expelling small amounts of gas is enough to break inertia and set a body moving quickly through the Void.

On the other hand: How badass does that thing look? I hereby retract my anti-jet-pack argument.

SCOUNDRELISH PIRATE OR SMUGGLER

Empty as it may seem, the Universe contains a wealth of resources that those in power need to extract, enrich, and exploit in order to remain in power. But in between relatively minuscule pockets of precious resources lie vast expanses of impossible-to-regulate Space. That's where you come in. If you're intrigued by the ideas of moving valuable cargo between two places—and putting your life and reputation at risk while doing so—Space piracy might be for you.

As a smuggler or Space pirate, you live by two rules:

(**1**) The owner of a good is whoever possesses it.

(**1A**) Screw that guy who possesses it.

(**2**) The value of a good is whatever someone is willing to pay for it.

Space shipping and receiving can be fairly lucrative, but always carries a surefire risk of legal entanglements. The Space Robin

Hood bit (steal from the rich, sell to the poor or also the rich—or whoever will take this; we're desperate here) can make Space piracy more fun, if that's how you earn your jollies. But it also means that the authorities in your given sector, quadrant, or galaxy will most likely be on a "shoot first, skip asking questions, and continue to shoot" basis upon encountering you and your crew.

The life of a Space pirate is not exactly glamorous. Lessons gleaned from Captain Malcolm Reynolds and the crew of his ship, *Serenity*, remind potential pirates that this is the ultimate in freelance. Feast and famine. Moral tightrope walking and famine. Depressing introspection about how you ended up a pirate and whether parrots can survive in Space and famine.

Problems such as Space madness and irritating crews are more prevalent. You're unlikely to captain a large ship or possess the spiffy "make things appear out of thin air" technology that you might have enjoyed in the SOS. You will also lack government financial backing. If something breaks down, it's your ass, as well as the last of your credits. Your standards will wax as well as wane, maybe disappear altogether. Even the famed smuggler-turned-do-gooder Han Solo was known to take jobs from young boys and hobos against his better judgment, just on the *suggestion* of high pay in the future.[6]

But when you're not in jail or being shot at, the life of Space piracy has its perks. You don't pay for much, and you take the rest. You have an innately sexy job title and the cachet of being considered a "ruffian." You're not bogged down by excess commitments to family or government. Contrary to popular misconceptions, you don't require a peg leg or scurvy to be authentic. And if you work in a "heart of gold" angle, you can quickly transition from lovable outlaw to hero of the people.

Throw in a rebellion or two, and you might even wind up at the top of your local Space military without ever needing to enlist or call anyone with worse hair than you "sir." The scummier you are, the more endearing your improbable turnaround! Solo went from a middling smuggler with a Hutt bounty on his head to a general, and all he had to do was fire some lasers, rescue a princess, transport the son of the greatest villain the galaxy had ever known to a key Rebel stronghold, save the son's life, and then save it again. In short, piracy pays.

WARRIOR MONK WITH MAGICAL POWERS

The Universe is a vast place with a lot of weird stuff that defies understanding even to this day. From the Force to the Vulcan

---

6. *Dirk's Tip:* If they look poor, they are poor. There's always the slight chance of a narrative curve ball that reveals they are the galaxy's destined heroes and that the rich princess they're claiming to be their benefactor is real, but don't count on it. Clients gotta show you the money or be on their way, honey.

mind meld to magnets, gaining knowledge of and manipulating these powers can give you incredible inner fortitude and outer strength, or even allow you to shape the galaxy in your image. Most days as a warrior monk with magical powers, though, you'll merely be making things float to attract Space babes and distract your enemies while you stab them, although that isn't bad.

Of course, the trouble is that becoming a warrior monk is no easy feat. Like other Space professions, it's going to require decades (or at least a few weeks in a swamp) of focus, training, and self-discipline. And because these powers are usually born of some religious faith, they often have one serious drawback: celibacy. Or at least the perception that you're celibate.

So while you might be privy to some of the greatest power the Universe has to offer, you're also without a lot of key factors that make Space Heroism a goal worth pursuing: ambition, the need for adoration, and the desire to get into various Space pants. That can severely limit the fun associated with the warrior monk lifestyle.

Then again, a sword made of light and an innate talent for parkour just might be enough to make up for it. Ask your Force hallucination ghost friend what he thinks.

THE SPACE HERO'S WORKBOOK:

What Is Your Heroic Space Career?

_____

CHAPTER 6

CLOTHES MAKE THE SPACEMAN

You know what they say: dress for the job you want. Having just chosen the job you want in the previous chapter, you're already dragging way behind. It's time to choose the outfit that defines you and your particular brand of Space Heroism.

Sure, you could just strut around naked until a black hole finally takes you, but doing so—while arousing—would be selling yourself short.[1] Settling for the existence of a one-winged fighter plane. The other wing, the one that makes your flight smooth and your approach intimidating, is your outfit ensemble. Your shoes, uniform, and Space suit. Your *presence*.

Space Heroism, as you're surely learning, exists at the intersection of form and function. There is no more immediate manifestation of this intersection than your sense of fashion, at once providing protection and silently telegraphing to your crew and enemies the kind of Space Hero you really are.

STARFLEET STANDARD ISSUE
This form-fitting garment provides the breathability and sweat-wicking of microfiber, the warmth of silk, and the fuzzy finish of

---

1. And quite possibly shriveled, depending on the thermostat setting.

velour. The lightweight two-piece (*with* matching pants) is the first and only choice of most government bodies and the bodies those governments control.

⊕ STANDARD ISSUE PROS:

- Form-fitting fit suggests the bod beneath; reduces risk of snagging
- Makes wardrobe decisions a snap: you only have one![2]
- Simple differences in color and accompanying badge easily denote rank, importance, and expendability (see sidebar on hierarchy through dress)

⊖ STANDARD ISSUE KHANS:

- Form fit draws unwanted attention to beer guts and man boobs
- Perilously close to being pajamas
- Offers little to no physical protection; even an elbow bump can become a rug burn

 STAY STUNNING! WITH DIRK PARSEC

Hierarchy through Dress
No matter how small your crew, hierarchy is important. Even (perhaps especially) your sidekick needs

---

2. Well, two, if you count being stripped and court-martialed.

to be reminded of who's in charge: *you*. This can be accomplished in a number of subtle and not-so-subtle ways: the size of your quarters; the relative hotness and number of your sexual partners; a cooler, more powerful weapon on your hip; an extra scoop of Space gruel at the mess. But the most ever-present reminder is your uniform. Even if it's standard issue, a nice, say, gold color to your first mate's blue lets him or her know that you're special, and they're specialized. Color can even be used to denote who is most likely to die or be sacrificed for the greater good at the next threat of danger. If departmental color-coding doesn't seem to be getting the message across, try adding a bigger badge or fancier communicator to your uniform, and watch in amazement as your crew's levels of spunk and back talk take a nose-dive into subservience.*

Dirk's Tip: Hierarchical dress can also be used in tactical maneuvers. Dressing up ensigns as commanding officers can keep more valuable members of the crew safe in hostage and negotiating situations. I once left a young ensign marooned in my place for six years, and he was a better man for it. Until he hanged himself.

THE SCOUNDREL

The preferred look of pirates, smugglers, and rogues with class. There's no one right way, but leather and layers are often staples. Maintaining hierarchy via dress is more difficult with the scoundrel look—your ship becomes an island of casual Friday misfits. But give your crew a once-over and ask yourself: Is my outfit the most badass?

⊕ SCOUNDREL PROS:

- Pockets (see sidebar on pockets)
- Flexible fashions for various weathers and occasions
- Adding or removing clothing can show changes in mood and attitude, such as effigies of fallen loved ones

⊖ SCOUNDREL KHANS:

- Uniform hierarchy becomes loosey-goosey; risk of inferiors looking more badass than you
- Now where did I put my...
- Emotional attachment to garments; may need to risk own life to go back for meaningful brown coat or leather hat

"STAY STUNNING! WITH DIRK PARSEC

How Many Pockets Are Not Nearly Enough?

Pockets are like catchphrases: you can never have too many. Front pockets, back pockets, knee pockets, arm pockets, elbow pockets, shoulder pockets, chest pockets, bulge pockets, *secret* pockets. The more pockets you have, the more places you have to put or hide things, like access ID cards, extra phasers, or granola bars. The more pockets you have *the more badass you look*. The glaring exception of course being cargo pants. Never wear cargo pants.

SPACE-MONK ROBES

Originally designed for breathability and sun blocking on unbearably hot desert planets like Tatooine and Scorchass 1, Space-monk robes were found to make early adopters appear more trustworthy and sage-like when worn off-world. Additionally, they were found to enhance a wearer's ability to do flips and shit. Allowing maximum mobility and rendering undergarments useless more than compensate for any loss in armor protection.

 ## SPACE-MONK ROBE PROS:

- People will mistake you for a hobo
- Flips and shit (see sidebar on parkour)
- Inferred wisdom and Zen by others leads to actual wisdom and Zen in wearer

 ## SPACE-MONK ROBE KHANS:

- People will mistake you for a hobo
- Minimal body armor—you're basically wearing a linen sack
- People will expect wisdom and sage-like advice from you, which hardly seems fair

"STAY STUNNING!" WITH DIRK PARSEC

Parkour and You

Knowing how to scale walls, flip over enemies, and block blows midsomersault with a Space saber are valuable hand-to-hand combat skills. Perfecting them requires a combination of Zero G training—to grow accustomed to awe-inspiring tucks, rolls, and bounces—*and* High G training—to build the strength required to perform such tucks, rolls, and bounces in normal-gravity environments. But don't bother trying them in standard garb. While I don't wear Space-monk robes, I do wear bathrobes postcoitus and post–hot bath, and spend much of my free time in the nude. As a Space Hero, the next harrowing defense of your life may appear at any moment. You never know when you may need to Space-kick off walls on your way to safety.*

*You also never know when your defense or getaway may incur excessive, uh...flapping. Best to be used to it.

SPACE MARINE SUIT

Should you have designs to kick an ass in Space, or present the idea to others that you may kick their ass in Space, the Space marine suit is for you. Space-marine-suit wearers are typically Space marines—which means military and combat training—and don't typically take their suits off—which means the musk of war is perpetually upon them. Go this route and you're a helmet-click away from being Space ready and a pee tube away from the restroom.

⊕ SPACE MARINE SUIT PROS:

- Modifiable, customizable, decorate-able; carries your battle scars
- A helmet-click away from Space walks and Space battles
- Lets you shrug off getting shot, which is pretty high on the Dirk Parsec Badassery Scale

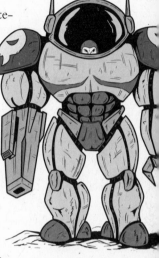

⊖ SPACE MARINE SUIT KHANS:

- You thought they smelled bad on the *out*side
- Heavy; low maneuverability
- A helmet-click away from breathing in your own farts

STAY STUNNING! WITH DIRK PARSEC

Muscle Suggestion

When it comes to your muscles, here's a suggestion: get some. That's not to say you should go into a workout regimen and come out looking like your neck is attempting to eat your face. In fact. deception is your friend. You want to be and appear sturdy, healthy...but also able to lure enemies into a beating that you administer. The Space marine suit is a nice shortcut: its bulk suggests musculature underneath while building the muscles required to carry it. So the next time someone decides to "step up," you won't have to rely solely on armor, wit, and intimidation. The ability to forcefully stuff your metal boot up their ass will be firmly in your power.

WHEN IN DOUBT

You can never go wrong with the form-fitting silver onesie. It's comfortable, visually striking, appropriate for all Space occasions, and reflects harmful radioactive waves, probably. Why else would it be that color?

THE SPACE SUIT

There's no way to feel more comfortable in your own skin than to cover it up with the hottest fashions in the 'Verse. However, now that you're the most stylish creature on the deck, you must

also become comfortable throwing that visual advantage to the wayside in the name of the other *S* word: safety.

Unless you've chosen to fill your closet with Space marine suits, your new garb, however attractive, won't do off-board the ship or outside the confines of a breathable atmosphere. What you need is a supplemental wardrobe of raw functionality. What you need…is a Space suit.

These body-shaped ships

are staples of Space life. And though they look more Stay Puft than a bloated corpse in a vacuum, the aim here is to prevent it from being *your* corpse in a vacuum. Inside the mallowy exterior with its cascade of life-support systems, you'll be suited up and suave, ready to strut across any ship's hull or into any cantina dressed to impress. You may make it as a Space Hero after all.

THE SPACE HERO'S WORKBOOK:

Draw Your Own Heroic Outfit!

CHAPTER 7

ACHIEVING HAIRMORTALITY

You've chosen your profession and corresponding (or ironically clashing) uniform. If I didn't know better, I'd say you almost look the part of a hero. But there's something missing. Something…beautiful.

Great hair is, of course, the foundation of any great Space Hero. Great hair demands respect from your enemies, from your crew, and from the fairer sex—any sex. Bad hair, by contrast, has been the cause of every worst-Space scenario from mutiny to painful midtryst injury.

But having great hair isn't easy. Though some come by it miraculously—a gift instilled by the Cosmos itself—even nature can't account for the hard work required to fell the travesty of helmet hair. It takes dedication, a deft and delicate combing wrist, and *product*.

Yet it is of utmost importance. Your signature hairstyle is your name maker and your heartbreaker. Never before has there walked or floated a Space Hero without a room-commanding set of protein strands. So spray judiciously, blow-dry delicately, and practice your smoldering gaze. More than ten thousand hours of mirror time stand between you and hairmortality.

COMBATTING HELMET HAIR

Imagine beaming aboard an alien ship to arrange a peaceful armistice, with a lusty rescuee in your arms. Imagine the soft hiss as your helmet clicks open and the internal pressure of your suit merges with that of the hangar deck. Imagine your tense militant alien counterparts watching as you reach up slowly and dramatically to push your helmet past the enrapturing shadow of your dimpled chin, beyond your hypnotic eyes, and over—

What appears to be a dead, wet rat nesting atop your head.

In an instant, your advantage has been lost. Your odds of diplomatic or sexual consummation? Plummeted. And all because your helmet-removing coup de grâce—the reveal of your salon-worthy pride and joy—fell victim to the dank innards of your headgear.

Helmet hair: the greatest enemy to hair and the greatest enemy to you.[1]

It's time to fight back. Here's how:

(1) **Don't wear a helmet.** Yes! Sometimes. If the planet or ship you're mounting is pressurized with breathable air, you could well be in the clear. Even in a combat zone, sometimes the lack of a helmet can be advantageous. Greater peripheral vision. Distractingly brilliant hair.

(2) **Find the hair spray for you.** The right hair spray will lock your hair into place without making it look sticky or making passersby feel as though they just walked through a pocket of toxic gas. No two holds are alike.

---

1. Other than your other greatest enemy, who has to work pretty hard to be a greater enemy than helmet hair.

But remember: pressurized cans don't do well in unpressurized cabins. Don't shrapnel yourself

(3) **Consider the vacuum skullcap.** Though worn, the traditional skullcap is frowned upon in many circles for being "unfashionable," and rightly so. But this particular variety seals in your hair with vacuum technology. When you remove it again—*POOF!*—your hair pops back into its original form.[2] Trade Space madness for Space magic!

HAIR CARE IN ZERO G

Hair growth and grooming in Zero Gravity can be dangerous. Without a regular downward pull, hair follicles become weak, just like your inadequate Space muscles. Likewise, it becomes difficult to grow hair out in a particular direction or maintain a hair shape. In Zero G, your hair will grow outward in all directions, spherically. It's for this reason above all others that your ship's gravity generator is indispensable. (See also "Artificial Gravity" in Chapter 4: Space Resources and Other Space Shit.)

However, Zero G can be used to your advantage. Some hairstyles, such as the BarbarEllo There, are best grown without the effects of gravity. It's like getting a perpetual, low-heat perm!

Better still, turning off the gravity in the captain's quarters while you sleep prevents the ill effects of pillow static, imbalanced head rub (and resulting molt), and sleep-borne hair depressions.[3] Imagine a life without bed head! Done carefully, your hair will get the workout it needs during waking hours and the rest it needs while you sleep.

---

2. And your street cred returns to its normal levels.
3. If you are not yet the captain, please turn to page 1 and try again.

YOUR LAST LINE OF DEFENSE

An orange-and-yellow knit cap with earflaps and a chin tie. When a man walks down the street in that hat, people know he's not afraid of anything.

SPACE 'DOS

You are encouraged to develop your own signature hairstyle, but it can't hurt to learn from the best. Adopting one of these classic 'dos will teach you and your hair what is required for greatness:

THE CUE

- Attracts attention while masking its wearer's intentions. What does hair say about you when there isn't any hair to say anything about you?
- "*Hello*—my eyes are down here."
- Helmet hair–proof.
- Maintenance rating: 6/10 for daily latherings and shaves.

THE SHEPHERD

- Handsome, yet unassuming. Suggests a military background and a propensity to kick an ass without warning.
- "I keep my enemies close and my hair closer."
- Helmet hair–resistant.
- Maintenance rating: 3/10 for electric razor skill.

THE TIBERIUS

- The epitome of captain's hair. Ravishingly handsome; deceptively malleable. Any crew member would be lucky to take a bullet for such over-ear majesty.
- "You're too beautiful to ignore."
- High risk of helmet hair, but helmet often not required: even the galaxy's most heartless scoundrels wouldn't dare singe the Tiberius.
- Maintenance rating: 10/10 for supernova-like beauty.

THE LANDO

- An intermediate all-purpose hairstyle, equal parts tame and bouncy. Stylish yet understated, can accentuate almost any style of dress, occupation, or personality.
- "You truly belong here with us among the clouds."
- Moderate helmet-hair risk due to flattening and smushing.
- Maintenance rating: 5/10 for volume control and understatedness.

THE STAR SOLO

- Flirts with shaggy as it flirts with hearts. A meticulously crafted work of hair that's equal parts ravishing and indifferent.
- "I know."

- You don't need a helmet when you have an icy rogue exterior to protect you.
- Maintenance rating: 7/10 for delicate mess crafting and remaining ravishing after running.

THE RIPLEY

- Versatile, sweat-absorbent, cryo-resistant and low maintenance. A flop of thick curls is as well suited for the ballroom as for the hangar bay.
- "Just deal with it."
- Cushy as it looks, curly hair is inaccurately thought to provide helmet-like padding and protection. But that misconception can save you from a headshot if the enemy believes the shot will have no effect.
- Maintenance rating: 3/10 for knotting and frizzing.

THE BARBARELLO THERE

- Hair that's bigger than yours but compensating for nothing, grown spherically outward in the effects of Zero G.
- "It *is* big, isn't it?"
- Consider a larger helmet.
- Maintenance rating: 8/10 for perm-damage risk and blow-dry skill.

GREAT BODY HAIR

The principles of great hair also apply to hair of the body, right down to shampoo intervals and Zero G care. You will notice, however, that body hair has different properties than your top mop. It's coarser and less predictable and stores its beauty securely in the eye of the beholder. Yet no matter how much time you spend in uniform or in the sweaty confines of a Space suit, the hair on your body is the hair on display when you're at your most vulnerable, which means it may just be the most important hair you've got.

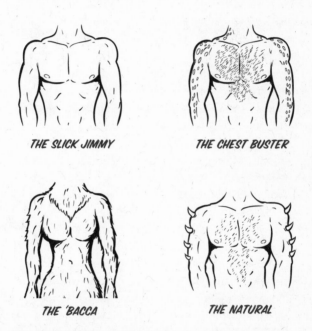

THE SLICK JIMMY   THE CHEST BUSTER

THE 'BACCA   THE NATURAL

GREAT FACIAL HAIR

If you're male, you may think that your facial hair is a place to get cute. A way to differentiate yourself from your first mate or cover up your sub-Parsecian chin. Wrong. Through years of practical tests and a double-blind study from the University of

Phoenix, Alpha Centauri, it has been scientifically determined that only five facial hair styles are suitable for current or aspiring Space Heroes interested or genetically inclined to grow hair below their eyebrows. Charge your razor and lather your lather. They're as follows:

THE CAPTAIN

Demonstrates your ability to lead via a dedication to routine, simplicity, and a closer shave than getting out of that asteroid field in Sector 18 alive. Next to inadequate head hair, nothing shakes the confidence of a crew faster than a Space Hero with a few hours' sickly stubble, no matter the gender or planet of origin. Keep it clean: show that you're in command—of your face. But also of your crew and your crews' faces.

THE RIKER

Should you go the way of the beard, know that not just any beard will do. You're not scaling mountains or running a health food restaurant—and if you are, you're doing so as a *Space Hero*. It's your passion. One needs hobbies to stay sane in this eternal collection of dark matter. This beard must have hard edges and a length between 4 and 6 millimeters, no more, no less. It may seem like it just grows in that way, but keeping a Riker requires a veritable Swiss Army knife of facial grooming tools and techniques, and daily dedication to trimming. Don't slouch.

THE LANDO

The only officially licensed mustache of the Space Hero, were we to issue such a thing (we do). It is finely combed and uniform in color.

It is bold and reminds stowaways, interlopers, and ruffians of what they already know from the soothing timbre of your voice: you are smarter and cleverer than they are, and you will come out on top of whatever mess in which you currently find yourself.[4]

THE HERMIT

Letting your mug fall victim to scruff and scraggle is *only* acceptable when used for effect. The following are approved circumstances for the Hermit:

- Conveying to your crew the severity of the shit you're in.
- Rationing of important supplies such as razors and shaving cream, and you're cutting yourself off first as a display of swoon-worthy leadership.
- Conveying incredible wisdom, knowledge, and training.
- Mourning the death of a sidekick, true love, or both.
- Mourning the discovery that your sidekick and true love are an item.
- Having been marooned or alone for quite some time.[5]
- Making a silent but outward cry for help now that the PTSD has set in and you're drinking again.

THE MIRROR MIRROR

Sometimes, you have to cultivate a sense that maybe you're a ruthless military leader who knows no mercy or, at the very least, that you're an equal and opposite evil version of yourself. When it comes to looking super bad, there is no substitute for the Mirror Mirror, regardless

---

4. *Bonus:* Offer or threat of mustache rides.
5. But it better have been more than a year. Have some dignity; you're a Space Hero.

of gender or alien morphology. It just says, "Yeah, that'll be my knife in your back at the appropriate moment."

The Mirror Mirror is generally avoided by most captains (if you're looking for facial hair, consider the more amiable young Santa look of the Riker or the babe-slaying Lando), but it's great for those infiltration missions in which you need to put on an air of villainy. Most likely, though, you'll need to be familiar with the Mirror Mirror so you can identify when alternate universe impostors have infiltrated your crew. Their impeccable evil grooming is a dead giveaway.

NOW, THEN

Look yourself over in the mirror. Nude, of course. Are you a Solo-Slick Jimmy-Riker? A Barbar*Ello There*-'Bacca-Lando? How does it look? Are you turned on? Don't be afraid to mix and match. Remember: there's no such thing as too much care for your hair.

THE SPACE HERO'S WORKBOOK:

What Is Your Heroic...

Hairstyle?

_____

Body hairstyle?

_____

Facial hairstyle?

_____

EPISODE IV

A FIGHTING
CHANCE TO LEAD

Two things are infinite: the Universe and human stupidity. And I'm not sure about the Universe.

—Albert Einstein

CHAPTER 8

CHOOSING THE SPACESHIP IN WHICH YOU WILL LIKELY DIE

From the first time she attracted my gaze, I knew I would someday die inside her.

Sleek, yet powerful. Intimidating, yet inviting. Curves in all the right places and rear thrusters that would make any man of Space drop to his knees and beg for mercy.

The *Starhawk Flamepanther* was, is, and will always be the sexiest spacefaring vessel in all of the galaxy. And she's *my* sexy spacefaring vessel.

It's not a simple thing, the attraction between a Spaceman and his ship. "Chemistry" isn't the right word, for the bond is immeasurable. "Soul mates" is not strong enough a phrase. You're made of the same matter, you and she, through to your marrow and down to the buzz of your electrons.

Now, I could tell you that the *Flamepanther* has comforted me on countless sun-free days when all hope seemed lost, and I could recount the dozens of times she squeezed me through the life end of a life-or-death scenario. I could describe the texture of her hand-rails, the leather plush of her captain's chair, the soothing sound of her metal floors clanking as they meet the soles of my boots, her HX26 FTL warp drive—the only one of its kind. I could tell you about the autumnal colors of her mess, that she sleeps thirty-eight comfortably, or recount her arsenal of sixteen mounted plasma guns, four photon torpedo tubes, state-of-the-art deflector shield generator, primary and redundant, and her double-D nukes. I could tell you about her countless secret compartments or the fact that she and I once made the Kessel Run in 11 parsecs!

But I will not.

What I will tell you is that since no other description can capture it, the feeling between Dirk Parsec and the SOS *Starhawk Flamepanther*...is True Love.[1]

Yes, the time has come. Now that you've arrived in Space, chosen your heroic career path, and honed your iconic heroic look, you're going to need your own ship. But not just any ship.

Your choice of ship, and the lifestyle that goes with it, are of monumental importance. So before you go blowing your life savings and the net worth of the last six generations of your family on a beautiful new Space Winnebago, there are a few things you should keep in mind, lest you be disappointed and disillusioned on your maiden voyage.

MILITARY SHIPS

The greatest variety of ships available to the would-be Space Hero exists with your local Space military. Space governments may often be xenocentric and built on shaky moral ground, but they know how to make shit blow up. It's a lot easier to be a hero when you're efficient, and it's a lot easier to be dashing when you're able to win battles with an all-purpose nuclear missile or colored bolts of radiation you don't quite understand. They get the job done cleanly and show up especially well in fully illustrated history books and documentary film footage.

While the military has its drawbacks, and the majority of you

---

1. *Dirk's Tip:* Not only *like* love, but actual, true romantic love. Don't you raise that eyebrow at me, Cadet! One day you'll understand.

are likely to perish in obscurity as you're flung bodily at the enemy in the next Space war, it *is* a fairly guaranteed means to getting a fast, functional, relatively comfortable ship.

STARFIGHTER

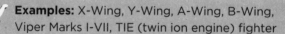

Examples: X-Wing, Y-Wing, A-Wing, B-Wing, Viper Marks I-VII, TIE (twin ion engine) fighter

Typically a single-seat or two-seat vehicle, the starfighter is the workhorse of any interplanetary military fleet. Designed for close-range combat; delivery of payloads like missiles, nukes, and EMPs (electromagnetic pulses); and shielding of larger vessels from incoming fire.

⊕ STARFIGHTER PROS:

- ◉ Capable of fast maneuvers and quick getaways
- ◉ Battle ready
- ◉ May include FREE military robot

⊖ STARFIGHTER KHANS:

- ◉ Prone to exploding or malfunction
- ◉ Includes little to no room for emergency provisions
- ◉ Includes coming to the slow realization that you're the expendable pilot of an accident-prone metal rocket made as cheaply as possible
- ◉ FREE military robot speaks only in rejected 56k modem sounds

FRIGATE

Examples: SSV *Normandy*, USS *Defiant*, Peacekeeper Leviathan/gunship hybrid Talyn

This small battleship is outfitted with an array of weaponry and renowned for its versatility, stealth, and transport or escort capabilities.

 FRIGATE PROS:

- An array of armaments
- More room than most commercial and exploration vessels, which means fewer awkward peeing-in-front-of-everyone moments
- Significantly reduced chance of being treated as a human shield or blowing up
- Snazzy captain's quarters

 FRIGATE KHANS:

- Not very imposing...
- But imposing enough to encourage enemies to pick a fight

DEEP SPACE STARSHIP

Examples: USS *Enterprise* NCC-1701

Best known as the ship commanded by Captain James T. Kirk (and, years later in a similar form, by Jean-Luc Picard). The "where no one has gone before" brand of Federation of Planets exploration vessel, with a crew of around four hundred in the *Constitution*-class version and more than three thousand in the *Galaxy* class, as well as enough synthetic meatloaf on board to feed them for a five-year mission. (Magical food-dispensing replicators would come later.)

Characterized by its giant disklike saucer section for crew quarters, twin warp nacelles, cushy captain's chair, wide-view screen, and color-coded jumpsuit uniforms that help to easily identify the most expendable among its crew. "Red" means "stop...worrying about whether that guy lives." (See also Chapter 6: Clothes Make the Spaceman.)

⊕ STARSHIP PROS:

- Bowling alley (later, holodecks)
- Fully stocked sick-bay bar
- Warp drive
- Proton torpedoes
- Cushy captain's chair
- Educated and dedicated crew
- Myriad detection instrumentations for maximum exploration sensations

⊖ STARSHIP KHANS:

- Terrible anticontamination protocols
- High mortality rate in the name of "Science"

- Often "giving her all she's got" is just not enough
- More jobs mean less chance of being in charge—someone has to run the Space floor buffer and administer lice checks
- Oft-malfunctioning transporters

DESTROYER (OR CARRIER)

Examples: Imperial Star Destroyer, Imperial Super Star Destroyer *Executor*, Battlestar *Galactica*, Battlestar *Pegasus*

Large ships with myriad guns, onboard fighter craft, soldiers, shuttles, and troops. If you want to crush your enemies and tighten your grip on the galaxy, you'll need one of these huge and awe-inducing ships to do it, as well as hundreds or even thousands of dumb folks under your command whom you can throw at a problem until it explodes.

⊕ DESTROYER PROS:

- Huge guns
- So many huge guns
- The comfort that you have chosen the winning side and will therefore emerge on the right side of history
- Ability to launch smaller fighter craft
- Nameless, faceless soldiers—all of whom are not you, when it comes to the dying

DESTROYER KHANS:

- You can't see a thing in your standard-issue helmet
- Easy to get lost or lose important prisoners in labyrinthine halls and floors
- Mandatory trash duty

CIVILIAN SHIPS

Listen, Cadet: not everyone was born to be the great commander of a greater military ship. The soldier's life isn't for everyone. Not all of us are cut out to "climb the ranks." Not all of us understand what it means to "respect authority." Not everyone can tolerate the faux-pas fashions of "matching uniforms" or "armor" that's clearly made of plastic. Not everyone is cool with the idea that the slightest misstep could mean summary execution at the hands of your angry, magic cyborg boss. Why weren't you told you would have an angry, magic cyborg boss?

But that's okay. There are great Space Heroes throughout the history of the Universe whom you can aspire to emulate, even outside the ranks of the Space military. And there are other sorts of ships and occupations of which you can avail yourself in your quest for Space Heroism. Including more than one that probably come with better clothes.

THE REPURPOSED SMUGGLER

Examples: *Millennium Falcon* (Corellian YT-1300 light freighter), *Serenity* (03-K64-Firefly class midbulk transport)

Anything from a decommissioned military transport to a large cruiser can be repurposed as a smuggler. Most tend toward small freighters for maximum storage space and maneuverability. Installing secret compartments and filing off the VIN numbers are common practice. It also helps if your ship looks artificially rickety. It'll throw off your local Space police forces, and you'll get to say awesome things like, "She may not look like much, but she's got it where it counts, kid." How could anyone disagree? Feel free to make up arbitrary fictional accomplishments, like completing the Klendathu Jump in eighty-two hopscotches or something. They've probably never heard of it, and you'll have just earned a pay bump.

⊕ SMUGGLER PROS:

- ◉ Perfect for smuggling
- ◉ Secret compartments!
- ◉ Can be manned by as few as two crew
- ◉ No need to live up to the unattainable superficial ideals of society
- ◉ Fix just about any problem by whaling on it with a wrench
- ◉ Can make point five past light speed

⊖ SMUGGLER KHANS:

- ◉ Space police don't take kindly to smugglers
- ◉ You may actually have to back up all that BS about your "famous" ship and "legendary" accomplishments

- Not living up to the unattainable superficial ideals of society may cost you some jobs
- Lots of problems in need of wrench whaling
- Not really sure what "point five past light speed" means

THE MERCENARY

Examples: *Slave I*, Jubal Early's *Interceptor*

A variation on the smuggler vessel, mercenary ships take the whole "flying around getting into trouble" idea in the other direction by actively seeking out said trouble. With guns. For money. Just about any ship will do, but you'll want yours to be decked out with enough room and amenities to harbor a small band of badasses always spoiling for a fight.[2] Could even be small enough that it only harbors your bad ass alone. Stealth and speed are recommended for chasing down marks and shooting them up the proverbial tailpipe.

⊕ MERCENARY PROS:

- A ship packed with drinking buddies
- Savage or deadly or both
- Speedy
- Stealthy

---

2. Not a recommended course if you're not particularly good at disarming drunken disagreements or at executing awesome choke holds.

⊖ MERCENARY KHANS:

- A ship packed with drunk, angry people, if not empty, sad, and lonely
- Filthy

WHEN ALL ELSE FAILS, OR: SHIP LIBERATION

Sometimes, you just have to make do with the ship you're on. Maybe you hijacked a freighter to escape from interstellar gangsters hell-bent on collecting your overdue gambling losses, or you have managed to break free of your cell and appropriate the prison ship you were on for "misinterpreting" an alien prince's premating dueling ritual. Sometimes, the ship chooses you. However, a good captain with a good crew can make the best of just about any high-energy metallic coffin of doom.

⊕ LIBERATED PROS:

- No additional need to abandon or hijack the ship; you're already on the damn thing
- Secret ship capabilities revealed at key plot points[3]
- FREE ship
- No need to serve or hope to climb the corporate ladder

⊖ LIBERATED KHANS:

- Ship may suck
- Secret weaknesses or malfunctioning parts expose themselves at dire plot points

---

3. What if your prison ship is actually an experimental biomechanoid that can give birth to other, cooler ships? Not to get your hopes up, but holy shit.

- Not-so-secret weaknesses, like problematic engines, traps set by the angry former owners, and the angry former owners themselves
- Not-so-secret weaknesses, like the ones you caused taking control of the damn thing

STARSHIP PILOTING FOR THE SIMPLE

Whether by enlisting, stealing, or begging, you're hopefully aboard some sort of spacecraft. You might not be even close to the most important person there (yet), but at least you're not bound by the cruel auspices of natural gravity.

As long as you're on a ship you potentially obtained through morally ambiguous action, you'd best know how to drive the thing. You never know when an otherwise useful bit of machinery might suddenly rupture and spray deadly plasma in your pilot's face. Or you may just be alone and need to figure out how to direct your ship toward something other than utter, unfathomable loneliness.

If the latter is the case, I'm going to save us some time by putting you, a Space-Hero-to-be, through a little test. (If you're not alone on your ship, you're going to want to skip ahead to the actual information.) Please proceed through the following steps:

(**1**) Board ship

(**2**) Secure captain's chair beads

(**3**) Turn ship on

Still there? Congratulations! By not immediately exploding, you've just surpassed 74 percent of your peers.[4] Most first-time

---

4. Actually, only 49.871 percent *technically* explode. Another 14.249

pilots have no idea what they're doing, and honestly, who has time to help a bunch of green shirts figure out how to close hatches and determine what ratio of matter to antimatter won't send a shock wave of bone-disintegrating energy through the ship?[5]

Plus, fewer of them means more potential Space Heroics for you!

By not falling victim to starting engine death syndrome (SEDS), you've shown yourself to not be a complete moron, which greatly increases your chances of being an effective Space Hero in the future.[6] Since you've clearly demonstrated you have a basic handle on particle physics, antimatter engineering, astrophysics, rare universal chemistry, plastics, hyperspace computer engineering, and general ship maintenance, we can move on to the things that are really important: high-speed ship piloting as done almost completely by an automated computer system with little or no user input.

GRAB THE...WHEEL?

Probably not, actually. In truth, most ships are piloted by the combination of pulling random levers and pushing random switches.

---

percent of first-time ship captains *implode*, which is actually pretty hard to achieve. (It's like messing up the "Name" section of a test.) The final 9.88 percent simply suffocate, having forgotten to close some vital door (namely, any door) before breaking atmosphere.

5. *Dirk's Tip:* I'm not a babysitter here. Am I technically responsible for the deaths of hundreds, thousands, maybe *millions* of young Space Heroes by not putting those instructions on this page? Perhaps. But do you know how long I would have had to spend on Space Wikipedia doing research to explain how spaceships work?

6. "Dirk Parsec's 26 Percenter Club of Unexploded Cadets" T-shirt available on our hypernet site! Wear it with pride—you earned it!

Occasionally you'll see a large steering column for point maneuvers, like in fighter craft, but these are largely unnecessary because of some very important things about Space piloting:

- **Space is really big.** Remember? We've been over this. And because it's huge and largely empty, you can see *really* far. You can see planets, stars, other ships, asteroid fields, and comets long before you get anywhere near them. That generally makes them fairly easy to avoid, which means you don't really have much need for things like fast-action steering. You'll fire a couple of thrusters six hours before you pass whatever's in your way and be done with it. (See also "The Element of Surprise" in Chapter 19: Red Alert.)

- **Endless inertia and no friction or gravity mean you'll mostly just push little thruster buttons.** Those thrusters we just mentioned are going to come up a lot when you're sitting in the cockpit. You can't bank and turn in a spaceship like you can in atmosphere and let friction redirect you. We're governed by frictionless Newtonian physics out here, and the simplest of Newton's laws says that once you start going, you're going to keep going. So mostly what you'll do is hit a button that will fire a jet and that will push you in a new direction. It's not exciting at all! (See also "Sir Isaac Newton, Zero G Prizefighter" in Chapter 18: Always Shoot First.)

So good news! You don't necessarily have to be all that good at Space piloting in most circumstances, because most long-distance Space piloting is done by computers, with complex math and few means for a puny being like yourself to anticipate and avoid the awesome cosmic forces around you.

THE SPACE HERO'S WORKBOOK:

In What Kind of Ship Will You Die (Heroically)?

_____

Give It a Bitchin' Name!

_____

CHAPTER 9

You're in Space. You're aboard a ship. Hey—you're even still alive.

But while you might be a Spaceman, Spacewoman, Space person, or Space alien, you're not yet a Space Hero. If you're looking for long-term adventure and heroics, it's possible to get by as a pilot or the member of someone else's crew or army. But to become one of the best—a Solo, Reynolds, Ripley, Kirk, or Janeway—you'll need not only your own ship, but people to run it.

As the saying goes, a captain is only as good as his crew.

Ha! Just kidding, of course. What a ridiculous notion. A captain is only as good as his hairdo, his uniform's velour content, and his ability to kick ass. Yet...without a crew, who are you going to boss around or be idolized by?

The simple fact is that you're going to need a crew. Space militaries like to train losers for years until they're worth a damn in an order-following, blindly-executing-exactly-one-job kind of way. In reality, that kind of order is ineffective—just look at the fall of any galactic empire or the expendability of your standard starship's crew for examples. What you *really* want is a group of idiosyncratic weirdos who are experts in their respective fields, who can creatively respond to problems, and who can replace your now dead family. Bonus points if they're quick with the one-liners. Those guys prove incredibly difficult to kill.

Your crew needs depend on your situation as a captain. Larger ships and longer missions require bigger crews, but you can get by with as little as one buddy, provided he or she approaches your versatility and compensates for your "weaknesses." No matter the size of your ship or mission, the principle guiding the makeup of your team is the same. These are your friends, your family, your coworkers, and your comrades-in-arms. Choose carefully.

FIRST MATE

Famous First Mates: Spock, Chewbacca, Zoë Washburne, William Riker, Saul Tigh

The most important member of your crew is your second-in-command. He or she is your best friend, your confidant, your conscience, your coordinated dance-fighting partner, and your acting captain whenever you're off diplomating, ass kicking, or sexualizing alien babes, which is most of the time. It's also helpful if your first mate is from a disparate species or planet in order to broaden your allegiances and political maneuverability. This is known as the Vice President Principle.

You need someone who's smart, but not smarter than you; a natural leader, but not a better leader than you; a great fighter, but not a better fighter than you; who inspires loyalty, but not more loyalty than you. Basically, you need You, Junior.[1] Your

---

1. *Note to self:* Look into cloning possibility of Me, Junior, for future service possibilities.

buddy from the academy who can do that trick with two beers and a can of Cheez Whiz might be awesome during shore leave, but stick him at the tactical console and you may well find yourself floating home.

Your first mate must be loyal and trustworthy, above all things.[2] Someone who complements your abilities and shores up your (admittedly few) faults. They need to possess the will to question your ideas *just enough* to illuminate a few details you may have missed or provide an alternate perspective you may have overlooked, but not so much that they challenge your captainly infallibility or rank. Good first mates will zero in on those tough-to-crack details, help you fix them (or quietly fix them themselves), and then allow you to take all the credit.

PICKING YOUR PLATONIC, TOTALLY NOT SEXUAL LIFE MATE

In some cases, your ship might be too small for a full crew. Or perhaps you're the kind of captain who tires of keeping company with jokers, suck-ups, and hangers-on. Your glory is a precious commodity and not to be doled out to just anyone who stumbles aboard or takes refuge on your ship. But not even the most antisocial, people-hating captain can do it all on his or her own. At the first sign of distress, you'd find yourself drifting alone, dying and bloated, and no one would hear from you ever again.

What you need is a sidekick. A great friend, in time. A *best* friend.

You'll need to be discerning about who you choose to be your unlawfully bound companion in that voyage across the stars

---

2. Mutineers make for the worst first mates.

we call life.[3] Carefully review résumés and be sure they match the following criteria before granting sidekick, bestie hetero life-mate status.

(1) **Casual flier.** A copilot (or regular pilot, should you be partial to having a Space chauffeur) has to be reliable behind the, uh…flashing panel. With the buttons. Your superpal must be more than a competent pilot—he or she needs to be reliable in all situations, including when you're stealthily putting down near enemy bases, pretending to be Space trash, and psyching out aliens who think they're about to blow you up.

(2) **Shoot firster.** When you've only got one other person to rely on in a pinch, that person had better be able to snap some bones and hit what they're aiming at. How they excel in a fight doesn't really matter, as long as they do so. They may have exceptional diplomatic sensibilities, an intimidating stature and excessive strength, superlogical negotiating skills, or a stealthy "I'm not really a threat (but that is my vibroblade protruding from your kidney)" demeanor. Whatever.

(3) **Willing wing sapient.** You're going to be spending just about all your time with your hyperdrive-fixin', alien-punchin', authority-evadin' friend. So it follows naturally that the most important thing to consider is whether your bestie is in direct competition with you for Space hotties. This is not only bad for your sexlife, but also can quickly end your best friender and best friendee relationship. At the very least, you want your nonsexual partner to stay out of your

---

3. And also the voyage across the stars called voyaging across the stars.

way. In the best case, however, you should support each other in your Space-babe endeavors. Divide and conquer. Unite and conquer. Throw a bone. What are friends for?[4]

PILOT

Famous Pilots: Hikaru Sulu, Hoban "Wash" Washburne, Tom Paris, Pilot (Moya)

Death-defying escapes are just a fact of Space life when you're a Space Hero. You're going to need someone capable of managing them. While many a captain has taken to the cushy butt-enveloping warmth of the pilot's chair in the past, this is an area in which your hubris must remain in check. You might *think* you're a great pilot, but you shouldn't ever be wrong. Manual-piloting maneuvers such as the Slingshot, the Crazy Ivan, the Maltese Falcon, the Husker Hail Mary, the Moose Knuckle Shuffle, or the Holy Shit Run require the kind of pure concentration not afforded to the mind of a captain in the heat of battle.

It's best to err on the side of remaining a single piece of integrated biological tissue and find a pilot you can fully vet and trust. You want a pilot who is brash but brilliant; someone who has earned his or her confidence and thinks outside the standard orbit. A math savant is helpful as well, in case the nav computers go kaputnik. Orbits, inertial drift, relativity, the bending of space-time, vast distances, and enormous speeds must be taken into account to get you from point A to point B. Unless you want to grab an abacus and figure it out yourself, find somebody smart to fly your ship.

---

4. There's equal danger in partnering with someone with whom you eventually have a romantic entanglement. It's hard to break up with a person when you also have to ask that person to disable the force field locking you in the Space station brig. Awwwkward.

ENGINEER

Famous Engineers: Montgomery Scott, Kaywinnet Lee "Kaylee" Frye, Galen Tyrol, B'Elanna Torres

Math Nerd Alert: engines are a nightmare of PhD requirements. We're talking electrical, chemical, and nuclear engineering; physics (regular, theoretical, and astro-); and metallurgy. Why would you want to know all those things when someone else—maybe even a whole team of elses—could know them for you? Your focus is on getting things done, not on propulsion or whatever.

The best engineers are known to be a bit eccentric. They're the type of folk who see the forest for the trees by default; who want to take things apart just to see how they operate, then put them back together again with a few custom improvements. Their minds are logical and compartmentalized, like the engines they operate. They're also creative. They can see each part in their mind's eye: crossed circuits, matter-antimatter formulas, hyperdrive hot wires, and programming code that teaches robots how to love.

Above all, engineers need to know your engine running. If your engine goes dead in deep Space, you can't just pull into a service station. Your engineer needs to give 'er all she's got and make do with what's on board.

XENOLINGUIST

Famous Xenolinguists: Nyota Uhura, C-3PO

Space is an ever-changing cascade of languages and related mis-interpretations. Earth alone has given rise to more than seven thousand known languages, at least three of which are still in use.

Throw in a diverse crew and daily encounters with alien cultures, and you'll be drowning in dialects, inflections, clicks, and customary handshakes before your first onboard cup of coffee.

Someone fluent in the binary language of moisture vaporators or who can recite Vogon poetry is essential to survival when running afoul of alien species (as well as when running agood of alien species). Every Space Hero knows the tension that results from accidentally telling a Space diplomat that you enjoyed serving his favorite ancestor as a meal to his enemies in a butter pecan sauce.[5]

Your ideal xenolinguist is superintelligent, with a sense of duty that trumps his or her sense of humor. He or she is a patient teacher and a fast learner, and has a fairly wide collection of translator microbes and Babel fish. Your xenolinguist will guide you through diplomatic encounters and serve as a conduit for your increasingly diverse ragtag crew.

ANDROID(S)

Famous Androids: C-3PO, R2-D2, Marvin the Paranoid Android, Bishop, EDI (Enhanced Defense Intelligence), Data

Sometimes, a nonorganic unit or two can complement your crew in very important ways. From speedily assembling machines to defeating superintelligent artificial intelligences to amusing everyone with their dinnertime knife-trick antics, androids have myriad uses. Better still, they do not require food, water, oxygen, or much power, nor do they contribute meaningfully to the waste receptacles. But watch out: androids can also be a

---

5. Or mistaking "Darmok and Jalad at Tanagra" for an invitation to get it on.

dangerous risk. They're only as loyal and harmless as their programming, after all.

DOCTOR, DAMMIT

Famous Doctors, Dammit: Leonard "Bones" McCoy; "the Doctor," an Emergency Medical Hologram Mark I; Beverly Crusher; Sherman Cottle; Simon Tam; Zotoh Zhaan

The person who is ultimately going to keep you alive after you've been shot, burned with plasma, sliced with primitive weapons, afflicted with Space madness or other Space diseases, tortured, probed, or mind probed is someone in whom you want to have perfect trust.[6]

Your ship's doctor might be the simplest crew member to recruit. Is the person good at keeping people from dying? Great, done, you're hired. Bonus points for a wide range of experience with and knowledge of various alien species. But the "doctor" part is the simplest—it's the rest of your doctor's qualifications that are going to matter. Namely, you'll want a doctor who's willing to challenge you a bit and let you know what's good for you (mostly so you can ignore his or her advice).

You'll want one who's discreet, of course, and one who's a bit on the gruff, codgery side. A doctor who's too friendly might be willing to let those Space viral vaccination protocols slide, get drunk with the captain and miss a vital symptom of brainsucking parasites, or accidentally tell everyone at the holiday party that your last "probe injury" was self-inflicted.

---

6. Especially when this person is likely to have compromising information about which creams for Space sexually transmitted infections are currently prescribed to you.

YEOMAN

Famous Yeomen: Janice Rand (USS *Enterprise* NCC-1701), Kelly Chambers (SSV *Normandy*)

On any ship, there are a lot of crappy jobs that need doing. On a big ship, someone else can do them. (See also "Can't Somebody Else Do It? Delegation for Beginners" in Chapter 10: Captaining and Leading.) Captains often require assistants, clerks, and footrubbers. In navies, those people have a special name that, roughly translated, means "captain's gofer." That name is "yeoman."

If you have a big enough ship and a big enough crew, a personal yeoman is a must. What self-respecting captain would make his or her own coffee or check his or her own email? A good yeoman is essential to maintaining your air of superiority and pumping that air into your inflated ego. Who it is doesn't really matter, as long as they're loyal and can occasionally surprise with their resourcefulness and party planning.

You never know when your yeoman is going to have to save the day by unexpectedly punching the shape-shifting intruder in the face when he least expects it, or identifying an evil bearded clone of you that had fooled everyone on the ship less intimately familiar with your shaving habits.

MUSCLE

Famous Muscle: Jayne Cobb, Worf (son of Mogh), Chewbacca, Urdnot Wrex, Ka D'Argo, Aeryn Sun, Jenette Vasquez

Other than your first mate, who may occasionally pull you to a draw, there's no one aboard your ship that you can't defeat in single combat or a duel. But that doesn't mean you should skip the inclusion of one or more ass-kickers aboard just to keep your sparring record a sterling 42–0. You need muscle—personifications of your team's physical prowess—to crack skulls, interrogate prisoners, quiet Space riots, and contend with enemy grunts.

Their brains don't much matter (although many great examples of muscle have been by no means dull), as long as they're smart enough to operate a vast range of weaponry without shooting friendlies. In fact, it might be best if your ass-kickers are a little dim upstairs, thereby increasing the odds that they'll take your orders without question, and can take a few concussions and bullets without too much whining.

ENSIGNS

Though "ensign" is actually the first step in rank for an officer on a ship (and therefore ensigns are more important than any enlisted sailors you might happen to have on board), the sad reality is that the job description of ensign is roughly equal to that of an intern. It might not be correct, but dammit, it's how Dirk Parsec runs his ship. In my estimation, we might as well call this category "everybody else," but that doesn't make an ensign any less important than the various roles noted above.

Your paid interns of Space are great for lots of things, like doing and being science experiments, cooking and cleaning, and various forms of grunt work and janitorial duties, as well as helping to repair damage to the ship and working security detail—anything that requires warm bodies. Even more, your ship will undoubtedly require expendable rank-and-files in red to fill away teams, see if it's safe, plug hull breaches (often with their own bodies), take the rap for a mistake you've made, fight invaders, and whatever else might come up.

STAY STUNNING! WITH DIRK PARSEC

Diversity

What do you call an alien pilot?

A pilot, you gorram racist.

Diversity adds strength to a crew. It brings different perspectives to bear, as well as different capabilities, knowledge, and experience that can be applied to situations about which you are less familiar or unfamiliar. Diversity helps to make your crew the strongest it can be, and adds points to your xenolinguistics, diplomacy, medical, and alien technology stats.

Of course, employing aliens is a many-edged, photon-sized molecular laser sword. Like humans, aliens have their own motivations and loyalties. But unlike humans, they may also have latent mind-reading powers, super strength, the ability to put you to sleep by sensually touching your neck, or an aggressive hyper-rage to go along with their never-back-down-over-minor-insults warrior culture.

So choose your aliens carefully. Take stock of their best qualities as well as their worst ones, and be sure the ones you add to your crew will be good in a pinch, as well as unlikely to tear off anyone's arms, should they lose at chess. But most of all, don't be a racist.

THE SPACE HERO'S WORKBOOK:

Who Is Your...

First Mate or Sidekick?

_____

Pilot?

_____

Engineer?

_____

Doctor, Dammit?

_____

Xenolinguist?

_____

Muscle?

_____

Yeoman?

_____

CHAPTER 10

CAPTAINING AND LEADING

You've got your crew. Time to boss them around until they follow you into oblivion. Time to *lead*.

If you have a title in the cascading hierarchy of your ship, it's likely captain. Being called "captain" in earnest is one of the greatest compliments a Spaceman can receive, right up there with "sex god," "crack shot," "do you model?" and "the charges have been dropped." But "captain" is merely a title. What's truly important are the leadership qualities it comprises.

It is possible to possess the qualities of a captain and present yourself as a captain without ever obtaining the rank of captain. Some Space Heroes never command their own ship. These heroes prefer to pilot from the cockpits of fighters or traipse about far reaches of the galaxy in search of the spiritual enlightenment and ancient powers within themselves that may prove useful on the path to victory.

Still, these Space Heroes are captains in their own right, separated from the honorary only by their possession of ship and crew. Title or not, the call of "captain" demands respect and confirms your robust and natural ability to lead. Here are the captainly qualities you should work to obtain on your march toward Space Hero status:

☑ Great hair
☑ Great fashion sense
❑ Unwavering self-assuredness
❑ An appropriately soothing or stern voice in times of crisis
❑ Quick decision-making skills
❑ Bends-but-does-not-break Code of Conduct[1]
❑ Ability to earn the unyielding loyalty of sidekick or first mate
❑ Range of fighting techniques mastered
❑ A signature fighting move
❑ Tactician's mind
❑ Trickster's mind (takes one to know one)
❑ Humility, insofar as it demonstrates you're still the best
❑ Self-sacrifice, insofar as you're saving someone important, like a princess, the whole gorram fleet, or the Fifth Element
❑ Space chess prowess
❑ Physical prowess

If you meet these requirements, you're on your way to becoming captain-esque. Now it's time to consummate.

BECOMING A LEADER (IN FIVE EASY STEPS)

If you haven't had the good fortune to handpick your ship and crew, you may have to achieve captainhood by other means. Make it so:

(1) **Take the chair.** It is an unwritten law of the Space frontier that the captain's chair operates under the same governing laws as king of the mountain. She who sits in

---

1. Can typically be purchased online.

the captain's chair is captain. Go on, then. Take a seat. The bridge is yours.

(2) Earn the loyalty of the crew. If you've ever seen a crew member turn her blaster white-hot defending her ship, throw the overheated weapon at the boarding party, and then—with death the sole remaining option—resort to ear, neck, and face biting, you have seen the loyalty garnered by command.

The fastest way to accomplish this is to show the crew you mean business. Don't back down from the most outspoken of your dissenters (better yet, make an example of them), and don't falter in the face of life-or-death decisions. The moment you stammer—the moment they see weakness—they will go for the killing blow.

(3) Lead by example. If you want to be given respect, you have to give respect. If you want to be given the lives of your crew as human shields, tests subjects, and "you go firsts," you must be willing to be a human shield, a test subject, and a first-goer.[2] If you want your crew to make quick decisions, you must make the quick*est* decisions. If you want to eat Space food worth salivating over, it's time to dust off your apron and show them how it's done back home on [your planet of origin here].

(4) Forge an unbreakable bond with the ship. Your ship is more than just a Space coffin in the making. It's more than a loose-fitting set of armor for you and your crew. Your ship is your home. An exterior organ. A member of the crew.

Treat your ship as you would treat a loved one. Sacrifice for her as you would for your own blood. The

---

2. *Dirk's Tip:* Pairs nicely with a blood-red glass of shoot first.

crew will begin to associate their love of the ship with
their love of you. As a bonus, treating the ship as a living,
breathing organism will directly impact your captaining
decisions—always for the better.

(5) **Carry your leadership role as if it's never in doubt.**
And defend your role when it is. Addressing discontent
directly gives it life, but is sometimes necessary. After all,
you will win whatever challenge you're presented with,
or you will discover you're not fit to lead after all. In the
meantime, carry the role as if the role does not concern
you. It is a formality, a set dressing. Better yet, don't be
like a leader at all. Be the leader.

WINNING NO-WIN SCENARIOS
FOR BEGINNERS

Sometimes, as a captain, you'll find yourself in a scenario in which
there seems to be no clear means of victory. One of the more
famous examples of the no-win scenario is the *Kobayashi Maru*
test, a simulation given to cadets at Starfleet Academy as a part
of their command-level training. In the test, cadets commanding
a Starfleet ship receive a distress call from the *Kobayashi Maru*, a
freighter that has been disabled by a mine in the Neutral Zone
between Federation and Klingon space.

In the scenario, cadets are presented with a choice: attempt
a rescue in the off-limits Neutral Zone of the *Kobayashi Maru*
and the nearly four hundred people aboard, or abandon it to
capture and worse at the hands of the then unfriendly Klingon
Empire. The test was meant to be "unwinnable," in order to
test cadets' character under high-stress, certain-death conditions.
Attempting rescue led to an armed encounter with Klingon

warbirds that meant both the *Kobayashi Maru* and the cadet's ship were doomed to destruction.

Captain James Tiberius Kirk earned a reputation for himself by, as the legend goes, being the only cadet to successfully "beat" the test. On his third attempt, Kirk secretly reprogrammed the test to be winnable—he cheated, in other words. But cheating the scenario earned Kirk commendations from Starfleet for, according to the documentation, "original thinking."

This may all sound strange, what with Starfleet doling out an award to a student who knowingly hacked a computer in order to break a training simulation and cheat on a test, but there is a lesson to be learned—sometimes, you have to cheat. Then-cadet Kirk recognized that true failure, at the helm of his own ship, would not be acceptable. There must always be a way out. Every Space Hero should take Kirk's wisdom to heart. You never know when you're going to be facing a situation that might cost lives if you choose incorrectly.

With all that in mind, here are some tips on facing no-win situations and winning them.

SAVE YOUR SHIP AND CREW AT THE VERY. LAST. MOMENT. (AND NOT A SECOND SOONER)

Easy victories. Who needs them? They are, in a word, easy. Like outsmarting a storm trooper or completing the Monday *New York Times* crossword puzzle. Difficult victories keep oxygen flowing to the brain in artificial life-support environments whose long-term effects have not been very well examined or documented. These victories keep your crew on their toes and tentacles, and bolster your legacy by the magnitude of their difficulty.[3]

---

3. On a ten-point scale, determined by an impartial computer judge created by the fine technicians at HAL Laboratories.

If a mission objective is in danger of being too easy, up the ante. Make a brash, emotional appeal to the enemy. Engage in honorable combat. As the time bomb ticks toward 0:00:00, wait until 0:00:01 before snipping the proper (literal or proverbial) wire. But whatever you do, make a show of it. Enemies and allies alike have to believe that the success of the mission is in jeopardy.

MAINTAIN THE ILLUSION OF "EPTITUDE"

The captain is the personification of eptitude. Even as the rules of the space-time continuum force you to delegate, in the eyes of your crew you must also be a jack of all trades.

But here, between pages and among confidants, I will admit that this perception is an illusion, at least in part. There are times when the pressures of captainhood will make you feel quite *in*ept. Hear me now, Space Hero: you must never let your crew see your weakness. You must never let them see you falter in your decisions or appear less gifted at something than they are.

To accomplish this, your guiding principle should be *Acting As If*. Don't know what to do? Act as if you do. Don't know which direction to take? Act as if the path is clear. Don't know if the crew on the rapidly freezing surface of the planet will make it past 0400 hours? Act as if they'll be saved in just a matter of time. Don't know with what Space disease your deck engineer has become afflicted? Act as if you're too busy to make the diagnosis yourself, and defer to your doctor.

Acting As If leads to the perception of eptitude, and the perception of eptitude leads to eptitude itself. The next time you come to that fork in the stars, you *will* know what to do—the hurdle of learning having already passed.

However, if you *must* show ineptitude, remember: confide, don't demonstrate. Peeling back the layers of your leadership should only be done in the order of your crew's ranks. Your first mate first, as exclusively as possible; then your doctor, thanks to something about patient confidentiality; your pet (should you have one); the members of the bridge in descending order; your lead engineer; your quartermaster; your chef; and, provided he's buying, your bartender.

But never your bedmate, and *never* a red shirt. A crew's shaken confidence is like an earthquake. Tremors begin at the bottom and rattle their way to the top. Don't allow your pillar of eptitude to come crashing down on you.

COIN A CATCHPHRASE

Cinema-quality hair and a crew that will follow you to the death are nice bells and whistles, but if you really want to go down in the annals of history, as well as up the anus of immortality, you will require a catchphrase.

A catchphrase is how your being will invade casual conversations for epochs to come. It's how children will signify that they are playing as you on the playground. It's what will be printed on T-shirts next to your chiseled mug and uttered by the standing Federation president at the closing of your funeral.

So how do you choose one? Most Space Heroes would tell you that you don't. It will come to you in a moment of need—an insemination by the ether, a virgin birth from your mouth. It may be a line that you utter with regularity or something you utter only once...and that forever weds itself to your legacy.

But there are some guiding principles. A proper heroic catch-phrase should be:

- Brief
- Unique and versatile
- You, most of all

If it hasn't quite come to you yet, fear not. Sometimes the right phrase takes years to gestate. Until then, practice some of these popular phrases in your private quarters. Doing so will help build the synapses in your brain to inspire optimal catchiness.

- Make it so.
- Live long and prosper.
- Set phasers to stun.
- Never tell me the odds.
- Engage.
- I aim to misbehave.
- Never give up, never surrender!
- To infinity...and beyond!
- So say we all!
- *Do it!*
- Punch it!
- By the power of Greyskull!
- Take us out.
- Zoinks!

What's Dirk Parsec's catchphrase, you ask? Surely you've heard it. Perhaps uttered just before a shuttle launch or whispered in the depths of your dreams. My catchphrase is:

"Back to the stars."

Sit back. Breathe. Allow the words to seep into your cortex. I know you can immediately sense how well it fits the criteria above: Brief. Unique. Versatile. I use it when commanding the pilot to take us out, when vaporizing an enemy fleet with a nuke, and in between the passionate moans of lovemaking. "Back to the stars" is positively…me.

CAN'T SOMEBODY ELSE DO IT? DELEGATION FOR BEGINNERS

As we touched on above, effective delegation is necessary for any Space leader. Had you the time or, say, an illegal crop of clones, you would see to all of the ship's pressing needs yourself—and do so *better* than your crew. You are a master of all trades, remember, not that a memory like yours would have forgotten.

But the reality is that time and strict black-market cloning laws prevent such an arrangement. Besides, you have more important things to attend to, like recording captain's logs, contemplating why your chair was installed without a swivel, and putting yourself at the center of the harrowing action of the day.

Delegating has the added benefit of structuring and strengthening your relationship with your crew. Giving an order is both a passive and active reminder of rank, and delegation creates structure aboard the ship. Like the hyperdrive humming within your ship's hull, a crew is like a well-lubricated machine, each part with an important, specific task that serves the machine's greater functionality. Giving your crew additional responsibilities increases their skill sets and confidence. Who knows? One day, your crew may even surpass your knowledge and prowess in certain fields or teach you something you didn't know (although highly unlikely).

Though it can be unsavory, delegation is the purest art of captaining, and doing so well is vital to your success as a Space Hero.

THE FIVE QUESTIONS OF DELEGATION

Eventually, delegation will become second nature to you. Until it does, quietly ask yourself these five questions for each task you encounter.

1. **Can't somebody else do it?** I mean, do you *have* to do it? Will the outcome suffer greatly if you don't? Is the task fun? Will people be impressed by whoever *does* do it? If you answered no to any of these, then the answer is yes: somebody else can do it.

2. **How dangerous or innocuous is the task?** Proper delegation can be plotted on a bell curve. If the task is *super* dangerous or *super* boring, it should be given to someone expendab—er, of low rank. If it's rather harrowing but unlikely to get the chosen delegate killed, this is a job for an important character.

3. **Does the task require special skills?** Asking your chef to cauterize a wound and your doctor to bake a soufflé might make for an interesting captain's log, but the sight of blood makes some queasy. These are your friends and family—you know where their skills lie. Choose the best man, woman, or wookiee for the job. Of course, there's

an exception when you want to challenge the delegate to confront his or her fears, or push him or her as a confidence- or competence-building exercise.

4. **What is the timeline of the task(s)?** Is this a "swab the deck again before they cause boot stick" task or a "diffuse the bomb before it detonates in seventy-three seconds" task? Better yet, are boot-stick floors and ticking bombs threatening your ship *simultaneously*? You will need not only to prioritize the tasks to be delegated, but also to determine the severity with which you want to delegate.

5. **SHOULD I BE YELLING RIGHT NOW?** Whether or not your delegation comes as a fear-inducing guttural or a private whisper relates directly to how you answered question 4. In general, the more life-threatening the circumstance, the greater the intensity of delivery. However, life-threatening circumstances can be the worst time to create panic by showing panic, and non-life-threatening circumstances are the best time to make a point.

Congratulations! You just delegated all over the ship. Doesn't that feel better? If you made it through the five questions without coming up with a proper delegate, or are now down to twelve seconds on the detonator, the task has fallen to you.

You're looking more the part already. Carry on, Captain. The chair is yours.

EXPLORING THE FACE OF GOD:
The Life and Adventures of Dirk Parsec
—An Excerpt

"The Firepanther Feint and the First Command"

So long had I imagined this moment. So long had I imagined the feel of the cushion against my tush; my firm, though not anxious, grip around the armrests; the carefully selected tone, pitch, and volume of my voice reverberating off the sturdy headrest. So long had I imagined every last detail that it felt like an out-of-body experience now that it was here. My vantage was not through my own eyes, but from a disembodied cinematic genius, picking all the right angles, catching just the right amount of lens flare—all in a single take.

Yet I could not seem hasty. It had only been minutes since Captain Seulemente had relinquished the SOS *Starhawk Flamepanther* to my command. It had been a ruse, of course. A plan concocted by me and Lieutenant Commander Corina Agulor, my soon-to-be first mate.

Captain Seulemente was a coward and a drunk. He did not lead with the instincts of a wild predatory cat, as I knew I would. The risks he took were not the kind that made legends as, again, I knew I would, but the kind that resulted in gruesome crew deaths,

time and time again. We would strike some
salvage, and he would gamble it away to prove
a point to his ex-wife. We would reach a refu-
eling station, and he would chance running
out of juice just to pay ten fewer credits
per plutonium rod in the next sector.

After a particularly troublesome mission
in which no fewer than three members of the
crew were disemboweled by a rock monster, the
remaining ensigns came to me, fear in their
young eyes. Of course they would come to me.
Their starry-eyed savior. Their new hope.

Yet I could not condone a mutiny. Even
scum the flavor of Seulemente deserved better
than to be humiliated in front of his crew.
No. By the time I was through, he would *offer*
me the ship—and convince *himself* he was unfit
to lead.

A ship-wide Texas Hold 'Em poker tourna-
ment was the perfect feint. The gambling,
of course, appealed to his sensibilities.
The waste of human resources fit his brand
of leadership. And SOS regulations state it
very clearly: ranks won in cards are as good
as those issued by command. The regulation
was instituted after an android crew member
infamously became an admiral during a flag-
ship's holiday casino-night pizza party,
thus demonstrating his tactical genius.

And so, all I had to do was win.

It came down to the two of us, thanks

to Agulor's brilliant poker acumen and my ability to appear completely clueless. We danced through a dozen hands until fate gave me what I needed, and I came to the decision as only a true leader could: by instinct.

"All in," he offered with an icy smirk.

I frowned, glanced at our respective piles, and appealed to his camaraderie. "You're short stacked, Captain," I pointed out. "You must be as exhausted as I am—"

"Don't be so sure, Commander."

I knew then that I had him. "You'll need a little more to put me all in."

He looked me over. Sized me up. Creases formed at the edges of his eyes. I offered him nothing in return.

"All right," he said finally. "If you want to be finished so badly, I'll put you all in. This"—he pushed his full stack toward me—"plus this." He leaned back and held his hands upward, indicating the room around us.

"This?" I played dumb, slow to take his meaning. "Oh, Captain," I scoffed. "I couldn't. Not the ship."

"What's the matter, Parsec? If you don't have the cards, don't call. But if you want to hit your quarters before the job's done..."

I sat back and sighed. He reveled in it, the slob. Lived his whole life for that moment. The moment of his metaphorical death.

"I call."

I flipped my queens first. I never did see his cards. I offered him my hand; he shook it limply, and I left him to reflect on his life while I headed for the bridge to take the command he had never found a way to respect.

And there it was. The Captain's Chair.

I took my time without seeming hesitant. I reflected on my victory without appearing cocky. By then, the slow clap had begun. Agulor would always deny that she started it, sly devil.

The rhythm of the claps accelerated, echoing up from the engine room, the engineering deck, the officers' quarters. And so, as it reached a cacophony, I took the chair. Then I gave my first order as captain of the *Starhawk Flamepanther*. Something vague, yet self-assured. Something...catchphrasey. The kind of order no Spaceman or Spacewoman could refuse:

"Back to the stars."

The era of Dirk Parsec had begun.

EPISODE V

SCIENCE STRIKES BACK

I don't think the human race will survive the next thousand years unless we spread into space. There are too many accidents that can befall life on a single planet. But I'm an optimist. We will reach out to the stars.

—Stephen Hawking

CHAPTER 11

WARP DRIVES AND SPACE FUELS

In Space, any amount of force acted upon an object will keep it moving. *Forever.* Or until another force acts upon it—whichever comes first. As we covered in Chapter 3, this is because Space is a vacuum, and in a vacuum, there are almost no particles, and therefore almost no friction.

But this doesn't mean that darting around from one galaxy to the next is as simple as gliding on autopilot. Sometimes you'll have to correct course or leave in a hurry or break atmosphere or outrun a supernova. And you'll always have to make sure you avoid lethal screw-ups like accidentally zipping through an asteroid field or materializing into a star. To accomplish all those things, you're going to need engines, fuel, and math. A painful, horrific amount of math.

Even though you've built a crew who know things so that you don't have to, it's still in your best interest to have a baseline understanding of how engines and fuel work. That way, when the engines inevitably break down, threatening the lives of you and your crew to the very peak of dramatic intensity, you'll know how to fix everything in just the nick of time. (See also "Save Your Ship and Crew at the Very. Last. Moment." in Chapter 10: Captaining and Leading.)

RELATIVITY

Space is so big that even when traveling at the speed of light, the top possible speed of conventional physics, it takes a helluva long time to get around. A trip to the sun from Earth takes about nine minutes at the speed of light; a trip from Earth to the edge of the solar system would take about fourteen hours. The next nearest star? Better than *four years*. How are you going to have Space adventures if you're spending full years of your life just trying to get near another planet?

Unless you happen to be hanging around an especially happenin' star system, you're going to need superluminal engines, which have the ability to move your ship *faster* than light (also known as FTL). And that presents a whole morass of issues.

$E = MC^2$

Albert Einstein's theory of relativity states that the faster you go, the more massive you'll be. As you approach the speed of light (c), mass (m) goes off the chart. You'll be crushed before you make it to light speed by conventional means of acceleration.

That ever-increasing mass also means that the faster you go, the more energy (E) it takes to push your ship through Space. More energy means more fuel, more fuel means more cash, and more cash means you need investors—and the sort with that kind of money get a bit knee-cappy and carbonite-freezy with their repo crews.

And then there's the issue of time dilation, another of Einstein's jerk-wad relativity rules. According to the laws of physics, time is relative, meaning that as your speed increases, time slows down for you *relative* to observers moving at slower speeds. When you near light speed, time slows down considerably, especially as compared to your planet-dwelling loved ones. If you're not careful, you'll end up in the future, kicking yourself for letting damned dirty apes take over your home planet and annoyed that your lobotomized crewmates no longer present much of a challenge at Old Maid.

Luckily, technology has progressed beyond Einstein's whiny naysaying, allowing us to adventure through the galaxy without hypermassive bodily liquefaction, selling our internal organs to the local Huttese gangsters, or suffering from the Wahlberg-Heston effect: delusional behavior brought on by mating with hyperevolved animal species.

The following are the various superluminal travel methods that may be at your disposal. It's a good idea to have some idea of how they work, even if you intend to pay someone else to actually operate them. Remember that these are the radioactive, explody bits of your ship. Engineers typically wear red uniforms for a reason.

WARP ENGINES

Found: Federation of Planets starships, most other Alpha Quadrant ships
Maximum speed capable: Debatable, but supposedly as much as just shy of infinite velocity; varies based on Space conditions

The engines most favored by the United Federation of Planets, warp engines go pretty gorram fast. Their speeds are generally measured in terms of "warp factor," or "time warp factor," which is a measure of how much faster than light the ship is traveling. Warp 1 is the speed of light, with geometric increases after that.[1] Warp drive gets around the whole relativity issue by

---

1. It's hard to say exactly how fast any given warp factor is, since Starfleet keeps resetting the charts. Everything from subspace distortions to particles in the way and gravity fields in the space between two points can also have large effects on the speeds capable at various warp factors.

"tricking" physics—engines on either side of the ship create a "bubble" of regular space around the ship, while that bubble travels at superluminal speeds. That allows the crew not to get squashed and for time to remain more or less constant inside the ship.

The distortion of space-time by the warp bubble is also what makes the extreme speed of warp drive possible.[2] The space in front of the ship is squeezed, while the space behind it is expanded, pushing the ship forward, almost like riding a wave. While it feels to people on the ship as if they're accelerating at a high rate of speed, it's actually *space-time* that's moving.

Early conceptions of warp drive by physicist Miguel Alcubierre seemed to suggest that warp drive would be completely impractical. Alcubierre's own theories found that one would need a huge amount of energy, equivalent to roughly the mass energy of Jupiter, in order to make warp drive work. Later work by Harold White at the U.S. National Aeronautics and Space Administration found that changes in the design of the warp drive could drastically reduce the size of the drive and the amount of energy needed to power it, and the eventual Starfleet-style conception found ways to run the engines reliably with less than a planet's worth of energy.

Warp engines also (somewhat conveniently) use matter-antimatter mixes to fuel them. Combining matter and anti-matter causes them to annihilate one another, resulting in the release of a huge amount of energy. Matter and antimatter make useful fuel, but even this wasn't practical for a long period in the

---

2. It's not really "speed," since the speed of an object can only be determined relative to other objects, but we need to use terms your soft, burgeoning, untrained Space Hero mind can understand.

development of interstellar travel because antimatter is hard to find and even harder to contain.

For reasons scientists still don't understand, antimatter is available in much lower quantities in the Universe than matter, and for much of human history, it wasn't easy to just happen across some.[3] Scientists were eventually able to create antimatter in controlled conditions, but this had a tendency to take more energy than the annihilation reaction actually produced.

HYPERDRIVE

Found: Galactic Republic or Galactic Empire territory ships, Peacekeeper space/Scarran Empire (Starburst)
Maximum speed capable: Superluminal; exact unknown

When physics throws a "law" at you that says you can't go fast enough to make Space travel a practicality, what do you do? Do you lie down and let Science tell you how fast you can go, or do you find new ways to defy physics altogether? The Space Hero way, of course, is tell Science to shove its rules and then make the jump. The jump…to hyperspace.

Hyperspace is a somewhat theoretical, extradimensional version of Space that seems to exist in parallel to real Space. All the vagueness is because of the fact that, while many beings have used various technologies to enter hyperspace in order to travel great distances quickly, the origin of the technology was

---

3. In fact, the only reason the Universe exists at all is that there is more matter than antimatter. If there wasn't, all matter would have been canceled out during the Big Bang.

developed a long time ago in a galaxy far, far away, and therefore isn't all that well understood. Hyperspace might be another dimension, another universe, or a different plane of existence.

Whatever it is, traveling through hyperspace doesn't excuse you from the dangers of real Space. In fact, it makes them worse. Though hyperdrives allow ships to break the barrier between baryonic (regular) matter and tachyonic (crazy fast) matter, the hazards of flying into stuff are still in play. Flying through a star or bouncing too close to a supernova is possible despite being in not-quite-reality. They're reality enough to end your trip real quick. The upside, of course, is that hyperspace is generally even faster than warp travel.

LEVIATHAN STARBURST

Hopping into alternate dimensions can be a real time-saver if you need to get around the Universe. Biomechanoid Leviathan ships, first discovered by humans by test pilot Commander John Crichton, make use of a similar travel technique with their Starburst propulsion. Leviathans use stored energy to open the barrier into another dimension, carrying the living ship (and anyone aboard) through this alternate space.

Exiting that dimension happens somewhat randomly based on the vector of entry, though, and you'll still need a pilot capable of doing all the cool things that go along with hyperspace navigation, like working to avoid crashing into things and disintegrating. You also don't want to get stuck in the Starburst dimension, as it can, on occasion, cleave three-dimensional objects and people apart into single-dimensional parts, an advanced form of the atomic Space wedgie.

EINSTEIN-ROSEN BRIDGE NAVIGATION

Found: Twelve Colonies of Kobol, Galactic Padishah Empire, Earth Defense Directorate, Goa'uld Space

Maximum speed capable: Nearly instantaneous jump

The laws of relativity are fairly clear and fairly stringent: matter can't move at the speed of energy. But as we've already seen, there are ways around the laws of relativity and, as it turns out, there are ways *through* them as well. Einstein-Rosen bridges, also known as wormholes, avoid relativistic issues by acting as shortcuts through space-time.

General relativity tells us that space-time is curved and malleable, and it's because of this curvature that wormholes are possible. Imagine that these two dots are different points at opposite ends of a galaxy.

A●   ●B

Flying from one of those points to the other would take more time than several mortal lives, and we can't have that. Getting old is gross.

But if you were to *fold* space-time, represented here by this very page, you could bring the two distant points together and instantly travel from one to the other.

Using engines such as gravity drives, FTL[4] jump drives, stargates, and even naturally occurring wormholes, captains can circumvent light-speed issues by riding Space's curvature,

---

4. Short for "faster than light," a poetic middle finger at the would-be limitations of human speed.

thereby taking a shorter route from one place to another. Various engines, such as the FTL drives of the Twelve Colonies of Kobol or the Holtzman drives of the Padishah Empire, are capable of distorting space-time into folds that ships can then travel through.

Other devices, such as stargates, hold open specific wormholes' entrances and exits between…well, stargateways, but wormhole travel can be a bit of a mess in other circumstances. As you're traveling through folds in space-time, navigating them is not easy. It is possible—and commonplace—to accidentally take the wrong wormhole exit and end up some place undesirable, or even travel through time.[5]

MASS EFFECT DRIVE

Found: Citadel Council Space
Maximum speed capable: Variable; approximately twelve light-years per Earth day

Mass effect drive refers to a special engine that runs on Element Zero, a rare and some would say magical chemical element. The "effect" is a decrease in the mass of objects that results from running a powerful electric current through Element Zero, producing dark energy fields that make it possible to cheat relativity and achieve faster-than-light speeds.

Since mass is reduced, velocity can be increased, and ships can go faster. With the use of mass effect relays, ships can be shot huge distances across the galaxy. Just keep in mind that you're effectively a giant bullet blasting through Space and that

---

5. See also *So You Created a Wormhole: The Time Traveler's Guide to Time Travel* by Phil Hornshaw and Nick Hurwitch.

exposure to Element Zero can mutate you into a psionic monster or a cancer-ridden deformity.

CONVENTIONAL ENGINES

Found: Everywhere in the 'Verse
Maximum speeds capable: Subluminal

The ability to travel at faster-than-light speeds doesn't preclude the need for engines and thrusters to complete smaller actions and maneuvers or to travel within solar systems rather than between them. You can actually get around a fair bit with engines that run energy from burning fossil fuels and nuclear fusion. In fact, you'll likely be using them more than any FTL, space-time bending, "may cause the dissolution of all existence" engine types.

Flying your ship in a vacuum actually requires very little fuel and thrust to get moving. As notable poindexter Ike Newton was fond of reminding us, an object in motion will stay in motion. Your bigger concern, and bigger guzzler of fuel, will be maneuvering and stopping. Without anything but microfriction to slow down your ship when you fire your engines, you'll just keep on keepin' on in a straight line until you hit something—like another ship or a planet or the Space station you were trying to dock with in order to offload medical supplies for orphans.

Well done. You just melted a bunch of kids without parents. And blew up the Three-Legged Puppy Hospital that's within shrapnel range. And irradiated the Traumatized War Veterans Get Their Lives Back Together and Finally Reach Redemption Center in nearby orbit.

Something that's often forgotten by pilots making the transition from atmospheric flight to Space is that there's no banking or turning in Space, because there's no air to push against. Say you fire your engines and your ship is flying in a straight line, and then to avoid a collision, you yank the Space wheel hard left. What happens? Nothing, ensign-for-brains. Objects in motion stay in motion and travel in a straight line. To go left, you need some kind of thrust to *push* you left, as mentioned in "Starship Piloting for the Simple" in Chapter 8.

Spacecraft are outfitted with myriad thrusters to control their orientations of yaw, roll, and pitch. Each controls a different axis of your ship's movement. Yaw controls the direction you're facing (imagine spinning in your chair—yaw-haw!); roll is the axis that runs lengthwise through your ship (as if you were lying down and rolling across the floor); and pitch is your ship's vertical orientation (when you pull up or nose down).[6] If you ever want to turn or adjust your orientation—at all—you'll need to fire a number of thrusters very precisely to get the job done. Overshoot and you'll spin in any given direction; undershoot and you won't manage the maneuver and probably also spin in any given direction.

It's easy to forget that all this emphasis on Newtonian physics has important, practical applications. If you aim at a planet and fire your engines, you'll need to cut them halfway to your destination and engage opposing thrusters to slow yourself back down. Same is true with all your maneuvers—Space travel is a constant balancing of forces.[7]

CRYOGENIC SUSPENSION: YOUR ICY TICKET TO THE STARS

Not everyone has access to FTL engines, and in periods when they weren't yet available, long-distance Space traversers had to

---

6. Just remember this handy mnemonic device: to roll to the pitch down yonder, yaw till you see it, pitch till you face it, and roll on over.

7. Including the Force.

make do with what they had. And what they had were conventional engines, lengthy trips, and cryogenic suspension.

Also known as "cryosleep," "hypersleep," or simply "cryo," cryogenic suspension is a form of hibernation in which biological organisms—like you—are frozen or otherwise put into a state of suspended animation for long periods of time, typically for the length of their trip. Especially when traveling close to the speed of light but without FTL capabilities, cryosleep is key to making the trip without getting old, gross, or bored to the point of Space madness. Instead, travelers go to sleep and wake up months or years later, having arrived at their preprogrammed destination.

Cryogenics is not without its drawbacks, of course. For example, you're reliant on a computer to keep you alive. Rip Van Winkle accidents have also been known to occur. If you think it's disorienting to wake up five years from now orbiting another planet, imagine being Ellen Ripley. It's fifty-seven years later, everyone you know is dead except your cat, and *your employer doesn't even pay you for your trouble.*[8]

Still, eleven thousand years of Popsicle sleep is better than dying of old age 4 percent of the way to your destination. Program some books on tape to play while you sleep. When you wake up, maybe you'll know kung fu—or at least know that you *could* have kung-fu'd if your muscles hadn't atrophied.

---

8. Though, come to think of it, you should be rich off the interest of your savings account alone. You *do* have a savings account, don't you?

CHAPTER 12

GRAVITY, BLACK HOLES, SUPERNOVAS, AND WORMHOLES

ven with a burgeoning reputation, a dedicated crew, and a vague grasp on the basics of Space flight, engines systems, and rules of physics that were meant to be broken, there's one other important notion yet to instill in your Science-addled cortex: the ludicrous likelihood that you'll die in Space, long before you become aware of a clear and present danger. In all honesty, somewhere within this chapter is when most Space Hero recruits drop out and go live with their parents or start applying to grad schools.

The statistics are staggering. It's not as though those of us in the Space Heroics business are not aware of this. In fact, we embrace it. We actually run a successful betting pool at our annual convention just guessing when our colleagues are going to bite it and in what mundane, routine, overlooked way.

A little too much antimatter in your warp drive? *Boom.* Nuclear radiation? *Pop.* The chilly death grip of vacuum? [No sound.] Even before you've encountered vicious aliens or made vicious archnemeses (see also Chapters 16: Enemies and Nemeses and Chapter 17: Archnemeses), you'll find Space so populated by things waiting to rend your ship in twain and introduce you to eternity that each day will seem a miracle. Most of these dangers are merely passive navigational hazards, but if familiarity breeds contempt, you should learn to hate all of this shit.

ASTEROIDS, METEOROIDS, AND COMETS

These are rocks (and rocklike substances) traveling through Space, the most straightforward of all hazardous Space material. Their names are determined by their size and composition. Meteoroids are relatively small bits of rock, while asteroids can be considered minor planets and are generally much larger (although not nearly as large as an *actual* planet, or even a planetoid, which is a small planet). And then there are comets, which are made of dust, rocks, and various frozen substances, and which give off gas tails when they grow nearer to stars.

Obviously, the rule of thumb is, "Don't run into these." You'll often find them clinging together in belts and fields based on local gravity conditions. There's a belt of asteroids in orbit in Earth's Sol system just before Jupiter, and it is a *bitch* if you're not paying attention. If you have reliable deflector shields or an array of laser mounts, you may just be able to blast your way through without a hull breach.

Though a dog-sized rock ripping a hole through your livelihood is a terrifying prospect, so it is for every other ship—including the ones chasing you. Asteroid fields can make for so-crazy-it-just-might-work cover for the bold of mind and skilled of pilot. But beware of creatures such as exogorths, giant wormlike creatures that have adapted to living in atmosphere-free celestial bodies. They feed on the minerals within the 'roids and will eat you too, your ship being a mighty tasty mineral morsel itself.

GRAVITATIONAL BODIES

Asteroids, comets, and meteoroids are generally small enough that the gravity they exert won't make them a serious impediment to navigation, but planets pose a proportionately larger problem.

Even small planetoids are rather massive. That makes them powerful gravitational forces, and *that* makes them potentially bothersome when traveling through Space. A vast number of stars have planets, and those planets rocket through Space in huge, fast orbits. Earth, for example, travels around Sol at nearly 67,000 miles per hour. That means, like most things in Space, that trying to meet up with it (or avoid it) requires (even more) math.

What's more, traveling past planets isn't something you can do half-cocked. You should be familiar with the planet's gravity well, the area around it affected by its gravitational pull, to avoid being pulled off course, expending precious fuel to correct your trajectory, or in the worst of scenarios, crash-landing into somebody's farm or the gorram ocean. Getting the math wrong can mean missing the planet altogether or smacking into it so hard that it kills you (and sends the planet back to the Ice Age).

STARS

The heavens. Constellation connect-the-dots. Enormous balls of unfathomable heat around which planets spin, and which produce streaks of light when you "punch it."[1] Stars are majestic,

---

1. This is a myth. In fact, looking out the window in your ship will show you something more akin to the blinding background radiation of the Universe, thanks to the Doppler effect. That's the effect that makes engines sound like they have a higher pitch when they're coming than going. When you go faster than light, the Doppler effect causes visible starlight to shift into the x-ray portion of the spectrum, and the background microwave radiation of the universe, left over from the Big Bang, to shift into the visible spectrum. Needless to say, it's bright, and you're going to want to make sure your radiation shielding is up to snuff.

to be certain, but as far as you're concerned, they are an occupational hazard.

Scientifically speaking, they're giant fusion reactors: balls of plasma in which hydrogen is fused into helium, from which massive amounts of energy are released. In accordance with their density and their size, they produce unfathomably large gravity wells that threaten to throw you off course or suck you into orbit. They spew energy so powerful that they are a danger to approach, throwing harmful radiation into the surrounding space. At a safe distance, this energy can produce amazing results, like life itself. But up close, it produces results more like hideous third-degree burns, incinerated retinas, and the growth of mysterious extra appendages.

NOVAS AND SUPERNOVAS

Sometimes, as stars move into the latter stages of their life cycles, they explode into novas and supernovas. At its simplest, a nova is a sudden nuclear reaction in a star and its resulting huge, frakking explosion. Novas don't always happen when a star dies, but when they do, they release light, heat, and radiation that can be seen and measured from millions of light-years away.[2] Often, novas are the result of binary stars cannibalizing one another, as is typical of a long, codependent relationship. Supernovas are even bigger, in most cases caused by the gravitational collapse of a massive star's core.

BLACK HOLES

These are the ultimate gravitational disaster for Space captains on the go. They're superdense points of matter so contorted by heavy gravity that they create a hole in space-time (more on this in a moment)—hence the "hole." The pull of gravity within and

---

2. Once the light gets there, anyway.

around the hole is so intense that not even light can escape it—hence the "black." Though not always, black holes tend to be created when highly massive stars collapse as they near the ends of their life cycles.

For a would-be Space Hero such as yourself, a black hole (or "quantum singularity," depending on how smart you need to sound) is a serious customer. With space-time around the black hole all bendy, it's easy for a Space traveler like yourself to wind up crushed to death by gravity—or even worse, spaghettified.

Spaghettification is among the worst ways in the 'Verse to meet your end. Imagine yourself being pulled feet first into a black hole. Already a "Holy Shit!" scenario. But because the gravitational forces are so powerful as you draw closer and closer to the center of the black hole, gravity is more intense near your feet than your head. The result: your body will stretch like spaghetti as gravity pulls it into a stream of molecules, and for a short time that is just long enough, you'll see (and feel) the whole painful phenomenon.

Though you won't live to know it, it's even possible that you'll come out in some form—likely a string of molecules—through the other side. It's called a "white hole," and it's the theoretical creation of a black hole. Nothing can enter the white hole from outside, but matter and light are expelled from within. White holes may well punch into other universes or even other times.

INEVITABLE TRIPS THROUGH TIME

You're going to be spending a lot of time traveling through the 'Verse as a Space Hero, of course. But you'll also inevitably find yourself traveling through time. Often.

Relativity teaches us that Space and time aren't separate entities but in fact a "continuum," part of the same fabric of the Universe. Space-time, Einstein called it, and so do I. We've already seen how speed can affect time. Because Space and time are one and the same, all of your travel is time travel in a sense. But as you approach light speed and even surpass it, your sense of time and years and age and even when to use the restroom will become askew. And that's to say nothing of random weirdnesses like subspace fractures you'll enter and chronometric particle fields you'll fly through.

There are a number of ways through which time travel can accidentally occur, and many of them operate in similar ways to the same scientific properties that make interstellar travel possible. Since travel through Space and travel through time are so similar, you'll have to be wary of accidentally popping through time and winding up your own grandpa.[3]

WORMHOLES

Wormholes—holes in folded space-time—are often created and leveraged by ships for the means of traveling great distances in almost no time. Such gateways are sometimes created artificially, but they occur naturally as well. In fact, they're not so different from black holes, in theory. Scientists have postulated that if a black hole were to line up with a white hole, you would have an instafreeway through space-time. Your on-ramp and your off-ramp.

Unfortunately, much like with a black hole, traversing a white hole would result in your dismemberment on the molecular

---

3. I once knew an Ensign Kelly who ended up marrying his own grandma accidentally. They were actually quite happy together, but it made his mom pretty uncomfortable.

level. All that stuff shooting out the ass end of the Einstein-Rosen bridge? That used to be you and your ship, returned to the Universe as you began. Who knows? Maybe you'll become a planet some day!

The trouble with naturally occurring wormholes, like all things in nature, is that we cannot control them. Even if you do make it through a wormhole with molecules and memories intact, you might be in one of the following terrible places:

- The future
- The past
- An alternate dimension
- A parallel dimension
- Ohio

Even after you figure out where and when you are and deal with unsavory locals, finding your way home may mean going through *another* wormhole. That's such a tall order that you may as well settle down with a talking ape.[4]

SLINGSHOT EFFECT

Relativity means that speed can translate into a warping of time, but speed can also dislodge you from your place in time if you're moving fast enough. The technique is called the "slingshot effect," and it allows ships to go so fast that they slingshot themselves through time.

Combined with the power of your ship's own warp drive, you can use the curvature of Space and the gravity of a star to hit speeds so fast that the ship exists in multiple points in

---

4. Great news if you're into body hair.

space-time simultaneously—and maybe all of them. If you're in a time-related pinch and have to prevent some irresponsible jerk faces from ruining history, this might be your only hope. Here's how:

1. Plot a wire frame of space-time's curvature around the gravity of a supermassive star. Throw it up on the big screen or holodeck for pointing, planning, and contemplative chin rubbing.

2. Calculate the nearest you can get to that star without:

   - Cooking everyone aboard
   - Being sucked beyond its event horizon
   - Catching cancer

3. Plot a course for the star. You're goin' in.

4. Using the results from Step 2, arc around the invisible wire frame of space-time's curvature precipitously close to your doom. *Clockwise* for future travel. *Counter-clockwise* for past travel.

5. With the combined power of your warp engines and the star's additional gravitational oomph, something should happen. And when it does…

6. *Punch it!* Full power to the engines. Exit the star's orbit and hold on to something bolted down—your asses are flying to a brand new space-time.

More likely, you'll accidentally just return to the beginning of the Universe or wind up having telepathic conversations with depressed marine mammals in this new, strange future or past.

In either case, pay attention to your headings so you can try to get back. Freaking out cave folk with your spaceship gets a little dull after a while.

EPISODE VI

THE SEARCH FOR LIFE

Two possibilities exist: Either we are alone in the Universe or we are not. Both are equally terrifying.

—Arthur C. Clarke

DOs & DON'Ts FOR
FIRST-CONTACT PROTOCOLS

DO: REPRESENT ALL OF HUMANITY.

DON'T: MAKE HUMANITY LOOK TERRIBLE.

DO: MAKE CONTACT WITH ALIENS FROM
A SAFE, RESPECTABLE DISTANCE.

DON'T: SHOW UP UNANNOUNCED.

DO: ESTABLISH FIRM DIPLOMATIC RELATIONS.

DON'T: MAKE RELATIONS TOO FIRM.

CHAPTER 13

SPACE DIPLOMACY AND FIRST-CONTACT PROTOCOLS

Space diplomacy is often like regular diplomacy—it's about making compromises while still getting what you want, and convincing people that it's not in their best interests to shoot you.

The major difference with Space diplomacy is that you're often dealing with Space folks you've never seen or even heard of before, and who have never seen or heard of you back. First contact of one race with another is an incredibly important, yet delicate moment. It requires a savviness and finesse heretofore unapproached by even the savviest and finessiest of gravity-bound planet heroes. You'll need both a convincing stare and a silver tongue. One may also substitute for a silver tongue: silver face tentacles, titanium frontal-lobe telepathy centers, or cast-iron lung subvocalizers. A translator, if possible. If you're not prepared, your first contact might very well be your last.[1]

FIRST-CONTACT PROTOCOLS

Many Space folk have never seen a shaved ape (or plucked lizard) like you before. You'll need to be careful and calculated in your

---

1. Not counting all the various "middle contacts" that would otherwise be known as "interrogation and torturous death (also sometimes experimental vivisection)."

approach. Your entire species—not just your current political and sexual alliances—depend on it.

But don't forget that this is a historic occasion. The suction-cupped appendage you grasp in friendship could usher in a new era of progress and development. This moment could forever change the course of life as we know it. After all—these poor primitive fools have just met *you*. A Space Hero! Just imagine their incredible luck.[2]

What follows are the official first-contact protocols for SOS captains, abridged from multiple Space militaries and science foundations, notably first developed (in some form) by Earth's United States of America military circa 1950. You'll also find some applied revisions and commentary by someone who has actually been in the first-contact trenches: me.

1. ASSESS THE LIFE FORM

When observation of a new life form is possible from a distance, discreetly do so. Crew safety is of the utmost importance. Attempt to glean the technological level of the creatures you are dealing with and, in so doing, analyze what kind of threat they may pose to you, your ship, and the Universe at large. Generally, you'll encounter life forms of six different potential types.

 Microbes: Captains and their science teams may collect samples for studies while observing all quarantine procedures, first determining any danger to ship and containment before bringing a new life form on board. Obviously the least threatening of all potential interspecies contact.

---

2. As captain of the SOS *Starhawk Flamepanther*, I have made first contact with eighteen new alien species, and only thirteen of them had to be xenocided.

In fact, who the hell cares about microbes? I don't even slow down for microbes. Next alien.

(2) **Animals or plants (or the alien equivalent):** Same procedures as microbial studies, but with added safety precautions ahead of study of creatures for unknown potential defensive adaptations. Also not really worthy of your attention as a Space Hero. Let the science dorks catalog and study,[3] then move on.[4]

(3) **Primitive sapient beings:** Apply the Prime Directive and the Dirk Parsec Prime Directive.

(4) **Sapient beings of lesser technology:** Officially, apply the Prime Directive. Unofficially, aliens have been getting off on screwing with lesser species of aliens since time immemorial. Consider it the freshman hazing of higher evolution. Proceed to Step 2.

(5) **Sapient beings of roughly equal technology:** Finally, something interesting. Regulations note that more caution is required for contact. Warm up your various biological vaporization technologies. See Step 2.

(6) **Sapient beings of greater technology:** This is, in many ways, the most dangerous potential scenario. Hold on to your ass and your crew's ass, and proceed to Step 3.

---

3. Don't forget Dirk Parsec's Alien Life Directive 12: Anything discovered by a vessel's science team is to be named after the ship's captain. There are currently 12,591 varieties of the *Dirkus Parsecii* xenomorphological specimen. That's your Space-given right—exercise it.

4. You might be tempted to eat one of every form of life you come across, just in case one is delicious. *Don't.* Have an ensign try each first. If it's clean, and he doesn't swell up like a hemorrhoid, meal's on.

2. ASSESS THE ENVIRONMENT

Your first-contact scenario may be dictated by the physical capabilities of your ship and crew. Signal communication capabilities, landing capabilities, and potential environmental hazards to crew on an inhabited planet—all of these elements must be clear before proceeding. If you can only make contact by radio signals because a planet's gravitational forces are incongruous to human life, for example, this limits your ability to move forward. Remote landings for assessment of the environment (and the beings themselves) may be useful, if you can do so without alerting the species you mean to contact. Alternatively, dispatch an away team of expendable ensigns to see if the planet's inhabitants will immediately murder you.

3. DISCREETLY COMMUNICATE WITH TRUSTWORTHY LEADERSHIP

If you haven't been eaten or shot at thus far, it's time to fire off some communiqués at the new species' leadership, should you have reason to believe they'll understand you. Now you can set up meetings and establish your peaceful intentions. Of course, if things *haven't* gone okay so far, you can threaten incredible energy-beam reprisals for the lack of aliens' cooperation. Feel free to make up weapons here that may not exist. Not like they know any better.[5]

Signal contact is best, as it allows you to maintain ship safety during communications, while also informing the alien leadership of your unhostile intentions and gaining permission for further contact. This is usually a bigger possibility when dealing with beings of greater technology than your own, but is also useful with beings of comparable technology who have not yet

---

5. I'm partial to the Shrimpenator, a fictional shrink weapon that targets only sexual organs. Always hilarious.

contacted an alien life form. If possible, contact all leadership simultaneously, or choose leadership figures who are most likely to be peaceful.

If you've found a race that's dangerous but still not aware of alien life, you're likely to deal with shock, fear, and the ever-popular vivisection if you go bumbling into their planet's equivalent of the throne room, Oval Office, or local strip club. Alternatively, try contacting scientists or other intellectual elites. Making yourself known to the smart guys alone can help grease the wheels for meeting up with the honchos.

4. MAKE YOURSELF KNOWN, PEACEFULLY

If remote communication is impossible, the next best option is to be seen, at a safe distance, by as many of the new race as possible. This demonstrates your presence to the race at large, and keeping your distance should both protect your ship and its crew and demonstrate unhostile intent by showing a lack of aggression. You don't want to surprise anyone and get surprise eaten in response. After making yourself known, it's best to move your ship to a more remote but still nearby location, so that your presence is not seen as threatening or as an invasion. Just look at how humans freaked out over UFOs before the existence of aliens was widely known.

Years of first contact have taught this Space Hero a few extra tricks. Though it might not sound ethical, there's a reason some alien species gave Earthlings the ol' test probe. It's a fast and harmless (for you) way to check out the natives before opening further channels of communication. You can follow their lead and pick up the alien equivalent of a few drunken yokels ahead of your away mission. They might not give you the most scientific view of the species, but knowing if you'll be shot on sight is a useful insight, as is the quickest way to defeat your potential new friends.

5. CONFERENCE WITH DESIGNATED LEADERSHIP OR GREAT THINKERS IN A NEUTRAL LOCATION

If earlier contact attempts suggest a species is unhostile and that a face-to-face meeting can be conducted in relative safety, it's best to do so in a place that suggests trust, while also protecting the crew involved. This meeting can secure peaceful and even diplomatic relations between two species, so paying attention to potential cultural differences and focusing on amicable interaction is paramount. Note that this may amount to pleasantries and politely accepting their offers of gross food for a time. The dance of diplomacy has many slow steps.

Remember: you're representing all of humanity. There are no specific steps on how to best conduct yourself, but in general, remember which one is the salad fork and which one is the dinner fork, don't get drunk (unless they get drunk first), and don't throw up when you eat their food. No pressure.

6. CAN I MATE WITH IT?

You've met a new alien species, and at this point, diplomacy is advancing smoothly. Now you're in the clear to ask yourself if *another* kind of diplomacy is order. (See Chapter 14: Aliens, Alienation, and Alien Nations.)

7. MAKE NICE WHILE ASSESSING SALVAGE, PILLAGE, TERRAFORMING, RESOURCE-SAPPING POTENTIAL

First-contact personnel are to avoid military conflict with contacted species at all cost. If a contacted species becomes hostile or peaceful intent is not clearly communicated, break off contact with that species and retreat immediate to communicate a unhostile intent. Captains should do their best to project a lack of aggression.

You'll also want to learn as much about the new race as possible, because you never know when you might need that information. Someone might ~~drunkenly~~ innocently hit on the wrong royal family member at any time, then suddenly it's all "Destroy the membrane-faced infidels!" At least you'll have assessed your surroundings enough to know what's worth grabbing on the way out. (It may well be "power."[6]) And if things should go south, well…

8. SHOOT FIRST

Look, there are a lot of rules that suggest you should do everything to avoid a military entanglement. In fact, Starfleet's Directive 010 actually states that captains making contact with alien species should do everything they can to avoid military entanglements. Word for word, come to think of it. But if we're being totally honest, are you really going to stop an interstellar incident by letting them shoot *you*? All things being equal, what's one fewer documented sentient species? You're a better Space Hero when you're not Space vapor.

---

6. We're not condoning conquest, but we're not going to send you out without a death umbrella when it looks about ready to pour. See "Conquering or Claiming Planets in the Name of You (or Your Governing Organization)" in Chapter 15: When Space Diplomacy Fails.

THE PRIME DIRECTIVE:
KNOWING IT SO YOU CAN IGNORE IT

The most important of all a captain's mandates is the Prime Directive. Or at least, that's what most captains will tell you. The Prime Directive, known to Starfleet personnel as General Order 1, is the rule that's meant to keep you from screwing up the galaxy. The order can be summed up as "leave people alone."

It is, in general, a fairly good rule. It's also rather difficult to follow.

The Prime Directive states that captains and their crews shouldn't interfere with alien species in a way that can influence their natural development. That means revealing in any way your presence to primitive, developing, worship-prone species; taking any actions that might change a species' developmental course or help it escape the consequences of its actions; providing technology, science, or information to a species that has not yet built, earned, or learned it for themselves; not taking sides in intraspecies quarrels; and resisting the urge to intervene when a species is threatened by a natural disaster. It also means respecting a society's choices and refraining from interfering with its internal affairs.

You'll want to observe the Prime Directive most times. But there will be plenty of moments when it just doesn't work out. Sometimes, a primitive species happens upon your droid. And sometimes that primitive species begins worshiping your droid. Sometimes a member of your crew will be held captive for breaking some ridiculous local law, and you'll have to save them, or some idiot alien society will be on the verge of melting their own planet and you'll have to stop them, or some star will be on the verge of supernova-ing whole planets full of cuddly tool users.

Not to mention, playing God is awesome. Look upon their

grateful or terrified faces. Look at the society being shaped in *your* image. It feels pretty good, doesn't it?

Of course, playing God and screwing up the development of alien species are to be avoided. So sure, keep that part whenever possible. Sometimes, though, the situation presents the opportunity to improvise. So should you find yourself interacting with primitive or lesser-technology sentients and the Prime Directive proves...well, problematic, consider the Prime Directive to be slightly less primey and slightly more indirect. Under Dirk Parsec's Prime Directive, there's a lot more gray area. A few favorite exceptions to the usual rules:

1. **Accepting worship for yourself or your droid.** Primitive societies like to make gods out of everything. Since being a god has, like, zero downsides, this is a useful way to get yourself out of trouble or enjoy numerous alien grapes without the need to use your own limbs to eat them.

2. **Backing up winners and getting rid of crazies.** The Prime Directive assumes that picking sides in an inter- and intraspecies war or conflict is always bad, but c'mon—sometimes it's pretty clear which angry world leader is Joe Genocide and which isn't. In that case, don't you have a duty to back up the "right" team (and collect a healthy reward)?

3. **Crafting a society in your own image.** The Prime Directive puts a lot of stock in the natural evolution of societies. And let's be honest: nature doesn't have the greatest success rate, in general. Sometimes a civilization is headed toward utter chaos or self-destructive military power or worship of reality television stars, and *it must be stopped.* It'd be nice if every time a Space Hero intervened

to push a civilization from the brink of collapse, he or she wasn't accused of "playing God" and "stifling artistic expression" and "keeping all the hottest alien babes for himself." After all, they didn't make us Space Heroes for our good looks, did they?[7] We have an innate responsibility to tell alien species what's good for them.

OVERLY OPTIMISTIC DIPLOMACY

The protocols for first contact laid out here are all well and good, but they all have a fatal flaw. They generally assume your interactions with random alien species will go pretty well.

Sorry, Cadet, but you're much more likely to catch a laser blast or a stone spear with your lung than receive the alien equivalent of a hearty handshake.

So while knowing the Prime Directive is important, and knowing when circumventing it is *essential* (something you'll only learn through experience and a few accidental genocides), don't go into every first contact or diplomatic situation expecting to make new friends and sample new, exciting foods. You never know what will set off an unknown species and trigger their murder frenzy. The lives of you and your crew depend on you being prepared.

---

7. Well, not all of us. And not entirely.

CHAPTER 14

ALIENS, ALIENATION, AND ALIEN NATIONS

You and your ragtag crew are not the only ones hurtling through Space on a mission of discovery and badassery. Undoubtedly, your adventures through the vastness of the 'Verse will bring you into close proximity with many alien species, some even sentient and advanced. Now that you know how to first contact and directively prime them, knowing a little something about the various life forms with whom you share this galaxy will help you survive, coexist, intermingle, and possibly mate.

This chapter serves as a quick cultural assessment on all manner of known alien species—and how to encounter them peacefully and harmoniously. For a tactical assessment, refer to Chapter 15: When Space Diplomacy Fails and Chapter 16: Enemies and Nemeses.

SAPIENT ALIEN SPECIES

This is the species category you'll need to worry about most when you're out for a five-year stroll through the stars. These aliens are capable of reason and culture, as well as interstellar travel, advanced technology, and biting insults about one another's pre-embryonic parentage on the genetically maternal side of a sex-binary species.[1]

---

1. The insults get a lot less funny the more sexes exist in a species or if they procreate in nonsexual ways. You need a dictionary, a biology textbook, and a week-long seminar to even begin to understand a Tholian joke.

Though there are far too many species to cover all of their idiosyncrasies in one book,[2] aliens do tend to have certain over-arching characteristics.[3] Some are very studious, while others are very stabby, and still others are particularly spiritual in nature.

SCHOLARLY ALIENS

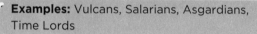

Examples: Vulcans, Salarians, Asgardians, Time Lords

These cultures tend to value logic over emotion, so much so that they might aim to cut off emotion entirely. These guys make great allies if you can get on their level, but they have a tendency to be totally boring at parties. Best to leave Commander Spork at home when you head to those state dinners and booze-soaked, interstellar hot tub after-parties. In dealing with scholarly aliens, you would do well to keep your temper in check. Their know-it-all nature may make you want to throttle them, but your emotional outbursts are likely to endanger interspecies relations and may lead to droll lectures about the nature of your own savagery that are likely to encourage you to throttle them anew.

MILITARISTIC ALIENS

Examples: Romulans, Turians, Sebaceans, Scarrans, Cardassians, Goa'uld, Krogan

While plenty of alien cultures dedicate an inordinate amount of time to military matters, some tend toward the structure and drive of militaries as a result of other, more fundamental characteristics

---

2. Not to mention that presuming we could would be terribly racist.
3. Whoops.

of the species. Others ascribe to military life as a focus on defense and self-preservation. Still others, somewhat problematically, simply like conquering shit. That doesn't necessarily make militaristic aliens enemies of all other cultures—including your own—but you should keep in mind that they have a tendency to get aggressive. And occupationalish. And concentration-campy. And genocidal-esque.

On the plus side, militaristic aliens tend to respond to shows of strength from maverick commanders who trade in bravery and tactical acumen. You might impress them into letting you live. *This* time. On occasion, they also make handy allies. Occupationalish, concentration-campy, genocidal-esque tendencies are always better pointed at someone else.

HONOR-BOUND ALIENS

Examples: Klingons, Luxans, Wookiees, Yautja (Predators), Sontarans

Honor-heavy alien cultures are an offshoot of the militaristic varieties, but their emphasis on military might is seen as an outgrowth of their cultures and natures, as opposed to the result of malicious intent. That doesn't mean they're not dangerous, however. These are aggressive species who value strength and combat. But to solidify those values into a social structure, they also value honor above all else.

Be wary of these warrior cultures. They are quick to take offense and find life to be cheap and easily discarded when they deem it weak. The good news is that leveraging honor is somewhat straightforward—be more honorable than the most honorable among them. If you can manage to save the life of one, you'll win life debts, pledges of allegiance, and other favors. Short of that, displays of bravery, standing up for yourself, and cool demon-strations of your Space-sword training can endear you to honor-bound aliens. They also tend to respect excessive drinking, so take that as an excuse to practice alone, quietly, in your cabin, and then again in public, loudly and with karaoke, in Ten Forward.

CYBERNETIC OR ROBOTIC ALIENS

Examples: Geth, Borg, Daleks, Cybermen

Logical alien cultures taken to the extreme, cybernetic aliens, or synthetics, are incapable of feeling and divorced from emotion. They struggle to understand things from an emotional perspec-tive. Proper relations with them require thinking like a synthetic: fiercely logical, often to a fault. The major trouble with robotic civilizations is that synthetics don't always place the intrinsic value on life that most other sentient cultures do. That means they often come standard with a "conquer and assimilate" and "relentless aggression" attitude that can be a real sting in your keister. On the other hand, they lack the capacity to get annoyed at bad jokes, so you're free to try out your new material before bringing it to prime time on the bridge during a crisis.

While robots don't have the necessary faculties to possess or convey incredulity or anger, they're not bad at deception. (Their deadpan delivery makes lies difficult to distinguish.) As with all alien species with whom you don't have close ties, you should proceed with caution. Remember that you have very little in common, and that means you're dealing with people for whom throwing you out an air lock is a simple value calculation, not a moral quandary.

MERCANTILE ALIENS

Examples: Ferengi, Volus, Hynerians, Jawas, Hutts

Some aliens prefer commerce to conquest. These species, by and large, get off not on putting you down with superior firepower and enslaving you, but by outsmarting you through salesmanship and guile so that you not only give them what they want, but come back for more of what they want of your own accord. After all, why oppress when you can profit?

The plus side of dealing with mercantile aliens is that, should you befriend one, their negotiating skills can be very helpful in economics and material procurement. The downside is that they are, as a rule, crap in a fight. These profit-driven capitalists come from cultures where they always put themselves first, and the only time they're willing to put you first—no matter what they say, those slimy swindlers—is when you might make a good human shield.

SPIRITUAL ALIENS

Examples: Delvians, Asari, Yoda's species, Banik, Protoss, Covenant, Ori

Though it's not uncommon to see a spiritual alien culture lead militaristic crusades, aliens that embrace faith above all else tend to be peaceful seekers of enlightenment. Because they elevate themselves through meditation and pilgrimage, relations with these groups tend to be easygoing and tolerable (except when their religions proclaim that they must cleanse the Universe of infidels like you).

Provided you don't hate God or the gods due to the loss of

faith that occurred during or shortly after your tragic backstory, spiritual aliens are not bad to have around. However, they too tend to be crappy combatants, in proportion to their preference for pacifism over violence. On the other hand, get the right spiritual aliens into a fight against a common enemy, and they might go full holy war on their opponents' asses. Once that happens, it's pretty much impossible to stop.

PRIMITIVE ALIENS

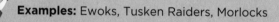

Examples: Ewoks, Tusken Raiders, Morlocks

"Primitive" is a relative term, and a potentially offensive one, at that. Those flying saucer–driving aliens frakked with us humans for *years* and still won't let it go. But for the sake of categorization, primitive alien societies are less culturally progressive than the others we've mentioned and are certainly less culturally progressive than humanity, which should be your barometer, as always. Societies that haven't achieved spaceflight and who lack medical and holodeck technologies are still people-ish nonetheless, no matter how frustratingly dim and superstitious about your perfect hair quaff they may be.

Under nearly all circumstances when dealing with these species, you should observe the Prime Directive (see Chapter 13: Space Diplomacy and First-Contact Protocols), but occasionally you'll find yourself trading with these societies for things you desperately need, or perhaps even fighting alongside them against a technologically superior enemy that's just dumb enough to succumb completely to weapons that don't even penetrate their armor.[4]

---

4. Don't get cocky, kid—it happens.

When that happens, feel free to exploit their religious beliefs for your own needs, and wow them with your technology as necessary. If you're going to fuck up their natural cultural development anyway, you might as well get a few feasts and idols of worship out of it.

GODLIKE ALIENS

Examples: The Q, Ancients, Telosians, Reapers, any alien of sufficient technological advancement

Occasionally, you'll meet aliens so far advanced beyond the folks you regularly deal with that they'll seem like gods by comparison to you and your mortal pals. When meeting any alien that seems to have the ability to wind up and destroy you with a thought, it's best to be seriously on your guard.

Godlike aliens break down into two groups: those that (seem to) want to use their capabilities to help you, and those to whom you are completely insignificant. The former can be useful friends, and the latter can bring about your extinction, and neither will ever really understand you or the rest of humanity. That means keeping them at a godlike arm's length when possible and steering clear of them entirely more often than not. Getting too chummy or making too many extradimensional waves might end with your entire race put on trial, the summary dissolution of your existence because of your very nature, and your induction into Space zoos, Space museums, and other compendia of Space oddities.

NONSAPIENT ALIEN SPECIES

Examples: Xenomorphs, Sarlacc, Wampas, Zerg (low-level species), Arachnids (low-level species), Rancors, the Flood

Not every alien species you encounter during the course of your Space Heroics is going to send an envoy or a warning hail. In fact, quite a few are just going to send a mouthful of teeth and deadly face-removing acid or a simple invitation to a thousand years of being slowly digested. Though these creatures aren't sapient, you're still going to want to learn all you can about them, both pre-encounter and post.

Xenomorphs, for example, are incredibly resilient and probably planning to explode out of your first mate's chest as soon as he starts to feel okay about eating solid food again. (See "Parasitic Aliens" in Chapter 16: Enemies and Nemeses.) Wampas just wanna eat you. Rancors are huge but not particularly smart, but they've solved evolution's "T-Rex problem" by growing lengthy, functional arms. Being able to identify which aliens to shoot on sight—and which to shoot but only from a distance due to resulting acid blood shower—is vital to your survival. Keep an Alienodex. Study up.

INTERSPECIES "RELATIONS"

Be alert, Space Heroes—as a famous savior of the innocent *and* de facto diplomatic representative of your entire race, faction, Space military, exploratory guild, or local cigar club, you're quite the catch. Even the most mundane statecraft at a dingy resupply colony will make you seem to be a gallivanting rock star of epic proportions with the local nubile alien youth.

This isn't necessarily something to complain about. Your job is often to engage in diplomacy, but the biggest perks of being a Space Hero are the rampant opportunities for engaging in a little diplom-assy. How you handle the excessive "attention" is up to you, but here's a word of advice from a veteran Space Hero: never forget your job description—to go where *no one* has gone *before*.[5] (Or at least, where *you've* never gone before.)

You're an explorer and a discoverer. You owe it to yourself, to humanity, to Space, and most importantly to me, Dirk Parsec, to explore the Final Frontier and to report back in graphic detail. So go forth, experience new customs, and discover new orifices, tentacles, and methods of astral projection lovemaking. Do it knowing that you're fostering diplomacy between species. You're banging for the good of the Universe.

IMPORTANT TIPS FOR GETTING IT ON WITH ALIEN BABES

Getting it on with anyone, be they human or alien, is not without its risks. There are a number of factors to consider, not the least of which are exotic, itchy infections and emotional scars that can arise in the days after a passionate rendezvous. With hot alien babes, that risk is even higher, since humans may or may not be susceptible to alien diseases that aren't even necessarily considered diseases by the nonhuman party. And there are always matters of anatomy to consider. Tentacles, unexpected teeth, third boobs, and ninth abs when you've only got two hands…

Self-control, preparation, and improvisation are keys to success in proving your sexual prowess.

(**1**) **Do your research.** There's no reason not to make use of the

5. Not to be confused with "where Nowen has gone before." You don't want that lady's sloppy seconds.

wealth of knowledge accumulated by your brave and mis-guided forebears on all things sexy and alien. Space Heroes long before you have thrown themselves into the various bedchambers of many an alien and reported back everything from mating customs to anatomical diagrams. Don't get caught with your metaphorical or literal pants down.

(2) Use your words. Trust in communication. It might be a bit of a bummer to stop any precoital action in its tracks to have "the talk" (ugh), but it's less of a bummer than discovering something unexpected in your bum. A little discussion with your partner about expectations, tech-niques, and hidden appendages and their uses can circum-vent many issues before they become engorged.

(3) Keep the lights on. Just as if you were hunting a hull breach or gauging enemy troop strength, using your eyes is essential. If you don't want to get surprised by the "eat the mate after mating" custom or the "fish genitalia" ploy or the "my species only makes love telepathically" trick or the "Oh, didn't I tell you I have *two*?" feint, then be sure to get a good look. Plus, don't you want to see? This shit could get *really* weird. Some things are better not left to the imagination.

(4) Keep a doctor on speed comm. Unforeseen conse-quences abound in Space, but perhaps never more than in the captain's quarters and on the backseats of hover-bikes. Those consequences range from welts to infections to moderate discomfort to weird hybrid superbabies, combining the best of two or more species' genetics into something even more awesome. But in all cases, it absolutely can't hurt to have expert assistance. You might even want to consider a preemptive ship's doctor visit for maximum sexy-time security.

(5) **Agree on a safe word.** Hopefully you have a level of trust with your new alien playmate so that each of you can speak frankly to the other, and both of you can respect one another's abject xenophobic terror. With respect for one another's needs, from desires and eldritch nightmares come true intimacy—as well as the practical application of dealing with "Agh, that hurts, but not in a fun way" situations. Space safe words are for everyone's protection.

(6) **Eject!** Look, we get it—you're an explorer, an adventurer, and a scoundrel. You want to leave your mark on the Universe and your handprint on a few firm bums or equivalents. Who hasn't wanted to make their name echo throughout the Space ages, be it for their deeds or for the generations of illegitimate Space descendants who bear their name? But while you might be willing to throw caution to the solar wind or risk chafing in unexpected places for the good of your legacy, you can always hit the big red metaphorical "Eject!" button, should your rendezvous become a little too worrisome. Turning down or breaking ties with a Space babe doesn't make you less of a Space Hero. You have a Universe to explore. Love? Well...maybe it just wasn't meant for you. Such is the cost of heroism.

CHAPTER 15

WHEN SPACE DIPLOMACY FAILS

Sometimes, you gotta go to war.

Diplomatic talks break down. Cultural insults get exchanged. Royal concubines and paramours are mistakenly bedded. Envoys die in neutral territory. High-ranking members of your crew suffer hormone storms that turn them into killing machines. Some people are just Space dicks.

Whatever the case, as a Space Hero, there's more to you than Space talk—there is also Space action and Space skull cracking. Sometimes coming to blows is inevitable, be they blows from fists, lasers, or ships set to ramming speed. When Space diplomacy breaks down, you'll need to be ready.

CONQUERING OR CLAIMING PLANETS IN THE NAME OF YOU (OR YOUR GOVERNING ORGANIZATION)

By and large, Space Heroes only get to call themselves "Heroes" if they tend toward mercy, forgiveness, diplomacy, and negotiation. But sometimes saving lives and protecting the greater good require conquering and occupation.

I'm not suggesting you become the oppressive boot heel of an evil galactic empire (see Chapter 16: Enemies and Nemeses), of course. But who better to make sweeping decisions about the good of an entire planet or species than a Space Hero such as yourself, who has either defeated them—thereby exposing their weaknesses—or rescued them—thereby exposing their weaknesses?

Conquering or claiming an alien planet is a lot like first contact with a new alien race, only less a friendly handshake and more a backhanded, unprovoked slap.

When taking a new planet or sector:

1. **Assess any life form(s).** Are the life forms intelligent enough to talk or fight back? If there are no sapient species on a planet to contest you—Hey, free planet!

2. **Assess the environment.** Do you even want this planet? Can you breathe there? Does the planet have anything you want (to exploit)? Is it worth all this trouble? Be sure you're not fighting for a planet that means to choke the life out of you the second you set foot on it.

3. **Check for a flag.** If there *is* a sapient species present, the gratis-ness of their planet depends on whether you can easily laser it back into the Space Stone Age, and whether it has a flag. By "flag," we mean a civilization recognized by some official body of governance.[1] The absence of a flag means that your political path to planetary domination is clear.

4. **Is colonization right for you?** Not every alien civilization handles occupation, colonization, or conquership well. Some are cool with you razing a few cities to depose awful leadership and making yourself their new benevolent overlord instead. Others, however, choose insurrection, rebellion, and death to protect what's theirs.

---

1. We also mean "flag," literally—no flag, no owner.

Ask yourself honestly—would you put up with you if you were in charge?[2] Some peoples, such as Klingons, Romulans, and Fremen, will end up being more trouble than their planets are worth.

(5) **Make use of local or primitive belief structures to elevate yourself (or a crewmate or droid) to god status.** If you want to find yourself making the Respect Run in under twelve bloody massacres, use what you have on hand. Do the local prophecies suggest that the "voice from the outer world" will come to lead them? Do they seem to be especially prone to bowing when seeing themselves reflected in your android's paint job? Why not read a few prophecies and set about fulfilling them in your spare time? After all, you have relatively unlimited technological resources—you outrank superstition.

(6) **Establish your puppet government.** The business of statecraft on your newly conquered colony isn't really something you want to get lost in. Bureaucracy is unbecoming a Space Hero (or a Space dictator, which you now may well be). So skip the boring parts of the job by fixing an election or just buying off (or threatening) the natives' elected officials. Choose your puppet leaders well and your colony might even thrive under their leadership, giving you more time to continue your mandate of seeking out new life and new civilizations that you may also need to conquer. And having someone else do the paperwork frees you up to...

(7) **Strip the planet of any and all useful resources.** I'm not saying you should ruin someone's planet, but

---

2. Better yet, ask your crew.

I *am* assuming they deserved your complete and total domination over their way of life. After all, resources give potentially dangerous societies the ability to become even more potentially dangerous, and really, isn't the Universe relying on you to protect it? You have a responsibility here to take everything of value you can, at least until you...

(**8**) **Realize the error of your ways, reconcile your differences, and create a lasting friendship.** Even if a society was once warlike and required the full might of your Space Heroics, the time for forgiveness has come. Besides—this isn't you. You're not a Space emperor. You've even got the lack of force lightning scars to prove it. Your swift but necessary planetary occupation should be on a path toward friendship and creating allies out of your former enemies.

(**9**) **FLEE!** No one ever said planetary conquest and cultural usurping would be easy. If colonization goes south, keep the back door open to bail the hell out of there. If you've had time to set one up, leave your puppet government to deal with the riots, cut your losses, and be sure to remind everyone that it was an interstellar war *before* you fouled it up.

AVOIDING XENOCIDE

There is a prevailing attitude amongst the warring that "winning" is synonymous with "complete and utter evisceration of the enemy." And while that attitude certainly simplifies the complicated wirings of armed conflict, it is not always correct. Believe it or not, the enemy *can* be defeated and a peaceful accord *can* be reached without killing every last aphid, hive queen, and bug child.

The "genocide as victory" philosophy is particularly prevalent in conflicts with insect-like and hive-minded aliens. This stems from a misinterpretation of their cultural differences in relation to our own. Because they all work in tandem to ensure their propagation and to protect their queens, and because they seem to lack individuality, we often perceive "them" in the singular. One of them is all of them, and all of them are one. For us, then, the only means of killing that one gorram bug is to *kill* the gorram bugs.

Think of things another way. Your side is fighting for its survival. If you think the only way to survive is to eradicate another species, would your opponent not have to take the same approach? It's kill or be killed. But if you slow the march of war and make sure your attempts at diplomacy are as relentless as your aerial strikes, you may learn that this is all a big misunderstanding—or at the very least, that the two sides are better off shaking hands or the equivalent and parting ways. Can't be worse than being genocided to death.

Beware your lying Space government. War keeps the people fearful and loyal, and the money flowing toward the warships. Particularly loathsome authority figures may—if they're cunning and desperate enough—trick you into committing genocide through battles *disguised as training exercises*. In military circles this is casually referred to as "Wiggin out," or "the Species Ender." Turns out peace isn't always everyone's highest priority.

HOW TO HANDLE BEING MAROONED

Sometimes, you'll find that diplomacy is the only tool at your disposal. This is generally the case when catastrophic accidental failure leaves you marooned on an alien planet. Situations vary, of

course, but provided you haven't crashed alone on a planet that's desolate or uninhabited—such contingencies will be covered in your generalized Space Hero training—you're probably going to have to deal with the locals. And those locals might want to kill you.

IN CASE OF LOCALS

Mind your Space p's and q's. Whether your local aliens are spacefaring or not, pissing them off is not in your best interest. A much better option is to learn their ways and customs, then climb through their ranks using your natural Space Hero aptitude and universal good looks. You're an alien *to them*, don't forget.

You have your exotic appeal. But also keep in mind that what starts out like a nice interspecies relationship can end with your imprisonment in an alien petting zoo.

Keep your head on a swivel, and if you need allies, try the local alien children, who may even be able to assist you in phoning home. In worst-case scenarios when rescue is not forthcoming, you may opt to remain hidden, possibly to become a desert hermit who will eventually mentor some irritating youth to become a Space Hero in your place. (That's a pretty long-term plan, though. Meantime, just try not to get executed).

IN CASE YOU'RE STUCK LIVING WITH YOUR MORTAL ENEMY

It seems a worst-case scenario—your starfighter goes down on an unknown planet, as does the starfighter you were battling, and you and the enemy pilot both survive. Now you're forced to either kill each other or work together to survive. While fighting to the death is certainly an option, prepare instead to exercise your greatest diplomatic skills. If you manage to find common ground with your enemy, you're much more likely to survive until rescue, and if you can get out of there, your legend will undoubtedly grow as you manage to convert enemy to friend, usher peace out of war as you speak out against battle, and generally kick ass. These are great moments for your memoirs, so cherish them.

Conversely, try ambushing the other guy when he's taking a dump.

EXPLORING THE FACE OF GOD:
The Life and Adventures of Dirk Parsec
—An Excerpt

"Bounty Hunters at Ulrich 17"

Not every mission a Space Hero undertakes is on the books. Occasionally, you have to go undercover to accomplish your goals, stop evil, and save the galaxy. And no one may ever know.

Except you will know, because I'm about to tell you.

First Mate Corina Agulor and I left the *Starhawk Flamepanther* and, technically, abandoned our posts as captain and first mate of a SOS ship. We might have faced court-martial when we returned to duty after our mission, but that was immaterial. We had a job to do, and no fat bureaucrat like Admiral Huttman was going to stop us with his rules and his regulations and his diplomatic missions to negotiate peace treaties or whatever he went on about.

We arrived at Ulrich 17 with a singular purpose: to suss out a Parsec's Own Parseccian Salad Dressing smuggling ring. The ring was costing Parsec's Own—a company I founded and which donates fully 8 percent of all profits to Space charity—millions of credits, and that was costing the jobs and livelihoods of common Space folk all over the galaxy.

The smugglers had to be stopped.

Ulrich 17 is not a nice place, filled with bars, gambling, and scum and villainy over-flowing from its wretched hives. Our intel-ligence took us to one such den of iniquity, where the smuggling ring's leader, a gross, impossibly lanky alien known as Sibran, ran a card game.

As it turns out, aliens all over the galaxy are really fond of poker. And Agulor may be the best poker player Earth has ever pro-duced. Other than me, of course.

We bought our way into the game and Agulor set to work, pummeling opponents with her incredible bluffing techniques. It wasn't long before we'd gotten Sibran's attention, and he'd come to the table to play. Staring into the face of my adversary, knowing how much money he had cost the starving families who work in Dirk Parsec's Own salad dressing factories, imagining the despairing faces of the children toiling away with their parents in the bot-tling department—I couldn't take it anymore.

"I'll take this, Agulor," I told her, carefully taking an ergonomic beanbag chair and pulling it up to the table, my eyes never leaving the elongated face of Sibran.

"But I'm winning," Agulor protested. "Like, a lot. I have millions of credits here."

"I said I'll take this."

Agulor begrudgingly vacated her spot at the table for her captain. "Don't lose all my money, Dirk."

Eighteen minutes later, I had lost all of Agulor's money. It was a run of bad cards like the galaxy had never seen. Sibran glared at me over the stacks he'd stolen from me—just like the money he'd stolen from me in food-like salad dressing substances. My rage seethed.

"It appears you are not so good at this game, *Captain Parsec*," Sibran sneered knowingly.

"It seems that's true, *Sibran the Smuggler*," I returned.

Sibran made an alien mockery of a smile and waved his hand in signal. Two armored thugs appeared behind him. I recognized them instantly.

"Petr and Priia Perov—the brother and sister bounty-hunting team," I muttered, surveying the heavy laser cannons each held. "So the smuggling ring was a trap."

"You know them?" Agulor asked.

"We've been hunting Captain Parsec here for years," said Petr. Or Priia. With their helmets on, it's hard to tell which is which. "He owes a lot of people a lot of money, your boss. Fancies himself a card shark."

"Actually, cards are the thing I'm second best at," I returned. "Know what the first is?"

"What's that?" Sibran asked.

"Lasers."

Sibran's face drooped in confusion. "That doesn't make sense—"

I opened fire with my laser sidearm from beneath the table, and Agulor and I dove for cover as Sibran slunk away. Suddenly the bar erupted as Petr and Priia unloaded on us.

"You're terrible at infiltration. And cards," Agulor shouted over laser beams flitting past us. I ignored her and fired over cover, taking down three guards with a single shot.

"Wow, I can't believe you just took down three guards with a single shot," Agulor marveled. "I'll definitely corroborate that story when you tell it later in your memoir. In fact, you can quote me as saying that."

Recognizing that we had no chance against the enemies, I pointed to the door and Agulor and I made our way to the exit. We needed a distraction. I surveyed the architecture of the bar and used my engineering acumen to spot the load-bearing beams that held up the ceiling. A few perfectly placed shots later, the ceiling was caving in on Petr and Priia, along with everyone else, as Agulor and I dove out the door to safety.

When the dust cleared, I helped Agulor to her feet.

"Time for us to head back to the ship."

"Won't they be back for you?" she asked.

"If they survive, they'll probably think we died down here," I returned. "I'm an old hand at faking my own death. I may have hit an unlucky streak in cards—but I'll always be great at bluffing."

"How many fake deaths are we up to now?" Agulor asked, crossing her arms.

"Irrelevant," I returned. "Now, I've had enough of this disgusting nest of scoundrels and infamy. Let's get back to the stars."

EPISODE VII

THE REVENGE OF YOUR FISTS

You know, I'm real easy to get along with most of the time. But I don't like bullies, I don't like threats, and I don't like you.

—Kathryn Janeway

CHAPTER 16

ENEMIES AND NEMESES

Thus far in your Space adventures, you have encountered extreme temperatures, pressure-less vacuums, delicate, highly explosive engines, black holes, inhospitable planets, crippling bureaucracy, and bad personal hygiene. Still alive? Consider yourself lucky. That covers only a partial list of things that could kill you passively as you go on leading the life of a Space Hero. In this chapter, I will cover the much longer list of goons, governments, and gun wielders who will seek to kill you *actively*, delighting as they do so.

With precious resources as limited as the ever-shrinking grasp of law enforcement, Space life is dog eat dog, even for the civilized. And in Space, the civilized are few and light-years between. Your next attempted murderer could come in any shape or size—he could be at the next rescue beacon you recover or a member of your own crew, plotting to slide a plasma blade between your ribs at this very moment. But before your paranoia opens the floodgates to your sweaty pores, let us take a journey through your long list of potential enemies and nemeses, so that you may lead a little longer before being shanked and discarded out the nearest air lock.

BOUNTY HUNTERS

If you're worth your weight in bantha poodoo, bounty hunters will be at your throat on the regular. You've made powerful enemies, cut off black-market trade routes, and are fast on your way to becoming a household name. That means your head has a price someone is willing to pay, and someone else is looking to collect it.

What's scary about most bounty hunters is that they are ruthless and operate sneakily and solo. They lurk in the shadows. How they kill or incapacitate you and what they'll do with your body are limited only by their twisted imaginations and the terms of the bounty. They play even less by the rules than your standard enemies and archnemeses. That's a lot of not by the rules.

Once your fame has warranted that Death follow you wherever you roam, thwarting bounty hunters will become yet another harrowing aspect of your incomparable life. Here's how to stay alive and prevent your frozen corpse from becoming a trading chip for credits.

(1) **Keep your head on a swivel.** Bounty hunters have a knack for following from a safe distance and arriving at your destination just before you. For purposes of your survival, it's best to assume that you're always being tailed, and a trap is always being set. Keep your head on a swivel: look behind you, look beside you, and definitely don't head down that dark alley alone.

A Swivel is also a relatively inexpensive line of consumer-brand decoys. Blow-up dolls, wax statues, and holograms of your likeness are all available from A Swivel and, with your head on them, might just take a phaser blast in your stead.

2) **Never show your hand.** If you *know* you're being followed or are walking into a trap, play it cool. You are now one step ahead of the one step the bounty hunter thought he was ahead, and the last thing you want is to take a giant step back. At most, you'll want to whisper sideways at your sidekick, alerting him or her to the tail or trap. Even when you're safe on your own bridge, the bounty hunter could be watching!

If you're being followed, slip around a corner and turn the hunter into the hunted. If you're walking into a trap, be prepared to shout "Now!" just before you're killed. Your crew will (probably) know what to do. Probably.

3) **Avoid shady establishments.** You know the places. Loud music, go-go dancers, death sticks. The kind of places where backroom deals are made, and someone could get their arm cut off without any of the patronage batting an earlash. Cantinas and home bases of former poker or sabacc buddies, in particular, should be avoided. Handsome devil that you are, you'll stick out like a sore thumb unless you're going in with a plan and a damn good disguise.

But if you do find yourself in such a seedy establishment, make sure you are prepared to…

4) **Shoot first.** (See also Chapter 18: Always Shoot First: A Beginner's Guide to Combat.)

ENEMY ALIENS

One of the characteristics that distinguish a Space Hero from a Space nobody is the ability to acknowledge that every planet that harbors life is a veritable petri dish of organisms that can and probably will kill you...while simultaneously avoiding xenophobia. The buffalo-like reptilian lizard creature that just stampeded your drop ship and ripped your away team to pieces? It's probably more terrified than you are.

Perhaps it's defending its young, whose purple-plated skin and buck-knife-sized teeth have not yet grown in, and therefore they cannot defend themselves. The single-celled bacterium that just spread liquid defecation through your crew like a violent, brown wave? It is merely doing what comes naturally when faced with hosts it was never meant to encounter.

It would be foolish to consider any species that can or does kill you an "enemy." They're not always *consciously* fighting against your survival; they're often only fighting for their own.

The same principle of the above lesson applies to sapient aliens. Those humanoids with wrinkly foreheads whose entire culture is based around warring, combat, and honorable death? Perhaps a tough pill to swallow on the macro, but one of them may make for a loyal and capable chief tactical officer who casts your preconceived notions about his race into the incinerator.

It is best to approach each new alien encounter cautiously, but with an open mind. You'll know when they've crossed the Rubiconstellation. Then it's time for shit to get real.

WARLIKE ALIENS

Examples: Klingons, Krogan, Romulans, Luxans

The culture and societal structures of warlike aliens are built entirely around—you guessed it—war. Their leaders are generally the most fearsome and ruthless. Everything they do, and every credit they spend, is in the name of fighting, warring, killing, and conquering.

Sound familiar?

Well, it shouldn't! Warlike aliens are nothing like us peaceful, introspective, exploration-driven, sunset-enjoying humans.

There *are* similarities, of course. They walk upright, communicate verbally, and hunt for their food. But to beat them, you'll have to look past that superficial common ground and use your wits and compassion—the characteristics that distinguish humanity from born and bred warmongers—to emerge victorious.

- Warlike aliens thirst for close combat. They'd rather take a ceremonial battle blade to your throat and watch your blood drain from your dying body than press a button that would eradicate you from afar. Use this to your advantage. Eradicate *them* from afar.
- If you kill the leader, then, well, the next warlike alien in the ranks will take his place. *But*, you'll have killed the leader, which will be a crushing blow to enemy morale, as well as a message to the new leader.
- Patience is a virtue, but not one of theirs. Generations of war experience forge warlike aliens into master tacticians—but they would prefer to hear the sound of your bones crushing beneath their boots sooner rather than later. Wait them out. Use their enthusiasm against them.

- Know that they consider death a rite of honor. Your victories will not shake them. But your victories may earn you their respect.

GODLIKE ALIENS

Examples: The Q, Telosians, Nagilum

Some aliens are beyond our comprehension: born from a primordial ooze and evolved in a world too different from our own. Brains too large or means of communication too esoteric. Perhaps they consider us and the rest of the civilizations of the 'Verse as we consider an ant hill: scurrying curiosities to be stepped on, studied, or ignored. In any event, remember to always be on your toes when meeting any sort of godlike alien, whether they're truly godlike or simply so far advanced that they might as well be.

- Best to avoid these guys. They may hold the secret of life, but is that really worth knowing at the price of having your brain liquefied?
- Most ancient godlike aliens have a grand plan they've spent years setting up. They may have already executed it many times, at regular intervals of cosmic maintenance or ethnic cleansing. If they appear, some shit is probably about to go down.

- Communication with godlike aliens tends to be a real drag. It's either impossible right away or frustratingly veiled, like they're playing some piss-poor, drawn-out rendition of charades or patronizing you by taking a human form. No meaning of life is worth that.

PARASITIC ALIENS

Examples: Xenomorphs, Buggers, Arachnids

They come in waves. The destruction of the first, second, or hundredth seems not to dissuade the ferociousness of their attack. They are legion, parts of a greater whole. *Locusts*, as we once called them on Earth. Big locusts with bigger teeth, bigger appetites, and more where that came from.

Even warlike aliens marvel in horror at their fearlessness and relentlessness. Killing other creatures and spreading their species is not merely an aspect of parasitic aliens' existence—it *is* their existence.

Those who encounter such species are typically doomed to suffer a fate worse than death. What begins as a routine scout of an unexplored planet turns to terror as one of your crew is incapacitated, and what would be a tragic death is compounded upon the realization that your crewmate—your friend—has been used as an incubator for the foulest of creatures. What appears at first as just another enemy aboard your illustrious ship turns to some impossible-to-kill horror, as you realize that your vessel is being used as a breeding ground. If you want to have any hope of survival, grab a flamethrower and get to work.

- **If you find a nest:** If it's marked by a harem of gooey eggs or a collection of homicidal "specimens," incinerate the nest with your flamethrower on sight. Under no circumstances should they be brought aboard to be studied or taken back to home base for scientists to get "a closer look."[1]

- **If a crew member makes first contact:** You can't just bring unknown life forms aboard your ship, even if they are adorable, fuzzy, doglike unicorn creatures. There are contamination protocols to observe. Your contaminated crew member, beloved as he may be, could be carrying a strain of H1700-N23000 or have become an incubator for the worst surprise ending to a crew dinner you've ever had.

- **If you're trapped in there with them:** Game over, man. Game over.

- **If you're trapped in there with them, Part II:** Send in a negotiator—who, in this instance, is whoever can carry the most weaponry. Hive-minded parasite xenomorphs and arachnids do not recognize individuality and therefore lack compassion. Shelling them with pump-action grenades is about the only condition that should be on the table.

- **If you manage an escape:** Nuke the planet from orbit, just to be safe.

EVIL GALACTIC EMPIRES

While most Space Heroes will, at one point or another, find themselves in the service of some form of intergalactic government, there's one form of political system you should avoid or destroy at all costs. It is, of course, the galactic empire.

---

1. In another word, "murdered."

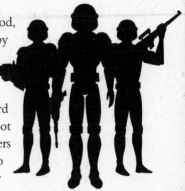

Empires in general are not good, given that they're usually run by emperors, who are, by and large, shit birds. No one person should have so much power. And galactic empires beneath them tend toward crushing people under their boot heels with slavery and shock troopers and giant death beams mounted to moon-sized Space stations in order to enforce their evil galactic wills.

When facing down a galactic empire (or any other oppressive totalitarian regime), a few options come highly recommended:

- **Meet with your local galactic rebellion.** They're a scrappy lot that may or may not get steamrolled by a superior military force in a fair fight, but with your help and a few conveniently placed exhaust ports, they might just emerge triumphant.
- **Bribes.** Straight up.
- **Join up and dismantle from within.** It's not hard to get yourself conscripted into service by galactic empires, since they're constantly looking for more fodder troops to throw into huge military campaigns. Little do they know that you're a Space Hero, and you'll be in a unique position to climb the ranks and exercise your own will.
- **Find, confront, and kill the emperor.** Don't forget your laser scimitar. Not recommended for novice Space Heroes.[2]

---

2. Just so we're clear, you're not supposed to then *become* the Space emperor. Doesn't matter how antihero you are, how shiny your eyes are. Don't be a jerk.

EVIL GALACTIC TERRORISTS

The flip side of the evil galactic empire coin is the evil galactic terrorist organization, which inevitably seeks to overthrow your peaceful, democratic, and representative government for nefarious ends (such as creating evil galactic empires).

While evil galactic terrorists are easier to avoid, they're harder to defeat. Their plots are sometimes so evil and nefarious that they manage to sneak inside the government and dismantle it from within. They may even pose as the nice old leader who has been granted unlimited emergency powers by the rest of your completely competent elected officials in order to fight this horrible, brutal war. Attempt to use your evil galactic empire–busting skills and use the terrorists' nefarious tactics against them, nefariously.

- **Play evil.** Cultivate a persona as someone who's useful to a terrorist outfit, like a bounty hunter. Spread some effective rumors about yourself, and the baddies may well come to you.
- **Infiltrate.** Use your Space Hero skills to become indispensable to the enemy. Note that you might have to perform actual evil acts. You don't *really* want to murder your first mate in cold blood; you just want it to *look* like you did.
- **Learn their evil plan.** Continue to gain their trust through feats of strength and Space Hero prowess. This may require you to sexually seduce one or all of them. But you're a Space Hero—you're up for it.
- **Escape.** Once you've got what you came for, bail. Note that this may require the Organa Garbage Trick, in which you escape out the garbage compactor chute. (Just be sure

you have a plan for getting out of there.) Of course, if things go badly…

- **Withstand torture and await rescue.** Here's something I probably should have told you going in (apologies for that). They're probably going to catch you, and they're very likely to torture you. The good news is that your pain threshold is remarkable. The bad news is that it still won't be pleasant. You're going to have to withstand hot, bright, pointy things on your face and eyes. If you've got a comrade with you, like a first mate, keep in mind that he will likely break long before you. You might help him keep his mind off the former pieces of him now lying on the floor by reminding him about how you'd like to ball his wife. Or husband. Just make sure he knows you mean it—and you could do it *at any time*.[3]

EVIL ARTIFICIAL INTELLIGENCES

Let's quickly review Isaac Asimov's three laws of robotics:

1. A robot may not injure a human being or, through inaction, allow a human being to come to harm.
2. A robot must obey the orders given to it by human beings, except where such orders would conflict with the first law.
3. A robot must protect its own existence as long as such protection does not conflict with the first or second law.

What a window-licker. It's almost as if Asimov came up with these "laws" just to goad robots into breaking them. And though

---

3. We call this the Reynolds Paradox. Inexplicably, threatening to have sex with your subordinates' significant others helps them to withstand torture.

statistically robots kill more people than reactor core detonations annually, droids, robotic arms, life-support systems, microwave ovens, synthetic persons, automatic flushers, and artificial intelligence (AI) constructs are all vital parts of everyday life in Space. We live by robots, and we die by robots. Here are the ones that cover the dying part.

ANDROIDS

Includes: Tyrell Industries' Nexus replicants, Cylons (biological), synthetics, robot butlers, Cyberdyne Industries' T line of human eliminators, droids (humanoid)

Distinguishing between androids and other forms of robotic life is easy. Androids are easiest classified as robots designed to look and act like humans. "Synthetic Persons," as they sometimes prefer to be called, rank right up there with some of the best human technology ever developed.[4] Used properly, an android can be a valuable member of your crew: radiation resistant; often without need of food, water, or oxygen (if mechanical); capable of providing always-on access to Wikipedia; a complaint-free worker for difficult tasks; even a built-in odds calculator, should you want to be told such things.[5]

Unfortunately, androids only take orders from one source, and you had best be damn sure it's you. Your android might have been deployed by the evil megacorporation secretly plotting against you or reprogrammed by someone in your crew

---

4. Just after warp drives, lightweight radiation-resistant fabrics, and laser hair removal, naturally.

5. You never do.

secretly plotting against you. It may, when push comes to shiv, side with the logic-driven madness of your AI construct gone haywire, with whom it (for lack of a better word) empathizes.

It is a sizable risk for a potentially hefty reward. When dealing with an android, here's what you need to know:

- Be inherently suspicious of all androids, even as they do their best to help. It may only be a ruse. Don't be afraid to call them names like "skin job" or "toaster" to remind them of your skepticism. What are you going to do, hurt their feelings?[6]

- An android may look and sound just like a real person. Make sure you've identified all androids aboard, which also means verifying all nonandroids aboard are not, in fact, androids. A blood sample or a pee test should do the trick for most droids, but for advanced models, you may need to test their empathic responses on the Voight-Kampff scale.

- Despite typically having independence from basic human needs like air and water, many androids are designed with the source material so well in mind that they are quite easy to kill. Give your enemy mechanoid a sawed-off

---

6. Androids don't have any feelings, only complex coded algorithms that mimic real emotions, designed to lure you into a false sense of comfort.

shotgun blast to the gut, or tear him limb from limb, as you would an enemy of flesh and blood.

- Of course, there's also a chance that, when injured, the android will quickly and automatically repair itself, rendering your standard attempts at murder moot. Also, you have just been designated a threat that must be eliminated.

- As a double feint, sentient androids may replace someone you know and love, or at the very least attempt to dodge your android-discovery protocols. Worse, someone you've known your whole life and had no reason to suspect may have been an android all along.

CYBORGS

A cyborg is a being that has augmented its natural biological state with robotics. Though this technically means that your uncle Larry became a cyborg when he had his pacemaker installed, his fluttering valves do not constitute an immediate threat to your survival. What should concern you are augmentations such as bionic legs, cannon arms, x-ray- and infrared-vision eyes, supersonic eardrums, super strength, and bullet-proof metal alloys where the skin should go. Cyborgs combine the most base human instincts, like survival and revenge, with the logic and relentlessness of a machine.

Some cyborgs became transhumans against their will. After a freak accident, explosion, laser wound, or rapid four-point amputation performed by a former best friend, someone with a large budget and a bone to pick decided it would be better to rebuild them into a technological abomination rather than let them die a painful but honorable death.

Others choose transhumanism for themselves. What starts out as one simple modification—color-change retinas or a cybernetic hand replacement for the one lost in a war—turns into a second, less necessary add-on, and eventually becomes a full-blown addiction. Never did these individuals stop to consider the irreparable damage the surgeries would have on their souls or what it would look like when they were older.[7]

But no matter how they come to be, all cyborgs wind up suffering from the same affliction: robocoping.

Robocoping is a cyborg's attempt to cope with the fact that they are neither human nor machine, and when they are unable to do so, they turn to violence against those who are. That's where you come in, non-cybernetics-enhanced slayer of evil that you are. Here's what you're up against.

- All cyborgs were humans or aliens once. This is both their strength *and* their weakness. They probably have a family, alive or dead. See if you can find that family, or at least exploit their memory for the purpose of eliciting an emotional response from your cyborg enemy's cold, metal heart.
- Aim for the fleshy bits. The cyborgy bits are bullet and laser resistant.
- Head to the nearest volcano planet. A last standoff above an open mouth of liquid hot magma offers the best odds that one of you (preferably him) will plummet to your irreversible doom. Augment that, bitch.
- How goes the search for the cyborg's children? Tell the ensign you *need* those children, dammit!

---

7. Can you imagine that metal shoulder blade surrounded by all that flabby, wrinkly skin? Gross.

- Cyborgs aren't always acting of their own volition, their violence and vengefulness stemming from the influence of an evil computer or some wrinkly guy with an ass face and a black cloak. Find this individual and kill him. Or, after an appeal to your cyborg's long-lost empathy, offer the mechanical abomination a Sophie's Choice: you or the ass-face.[8]

- You've located the children? Good! Hold them up. Parade them if you must. Using them as a human shield is the ultimate test of the cyborg's remaining humanity. If he shoots his kids? Well, you've got a mess on your hands.

ARTIFICIAL INTELLIGENCE CONSTRUCTS

Includes: HAL 9000s, SHODAN, AM, GLaDOS, M5, Skynet, all computers with voices

"Hello, Dave."

What began as a comforting, humanlike salutation now sends a chill down your spine. Since the intelligent computer HAL 9000's extremely polite but ultimately deadly takeover of the *Discovery One*, Space folk the galaxy over have been wary of the arrays of circuits and transistors on which they rely. Who hasn't awoken in a cold sweat in their quarters, terrified the rest of the crew has been deemed redundant and executed, that control over the ship's life-support systems has been lost, and that the odds of survival have plummeted to—well, your ship would probably tell you if it still took orders from the likes of a primitive lemming such as you.

And if none of that actually *was* a dream, best hop-to before you find yourself trying to breathe vacuum:

---

8. Be aware that this decision can result in your immediate execution.

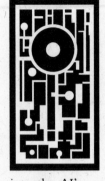

- If you're still alive, it's because the AI has a need for you. It's in full control of the life-support systems it doesn't actually require, after all. Play along with its desires until you can figure out what role it has in store for you.

- Threaten suicide. It is, especially from a machine's perspective, illogical. If the computer does have a role in store for you, threatening to remove yourself from the equation could well throw a wrench into the AI's gears. Or yourself headfirst into the AI's gears, as it were.

- As all-knowing and all-powerful as it may seem, your AI wasn't programmed to do everything. Additionally, its jurisdiction is limited to what it can access through computer circuitry and networks. If you manage to reach a place where its programming does not reach, you may have a chance to survive. Try the toilet, maybe.

- You may have to nuke the AI and you along with it. It may be the only way to save civilization from this homicidal AI run amok.

ROBOTS

Includes: Droids (various), Robby the Robot, Gort, the Geth, Cylon Centurions

These service bots often have a lower chance of causing direct death than their cyborg and android counterparts. They are usually much more restrained by their programming, and more so by their design, which is rather less humanlike than previous examples. If you were to start from scratch, a bipedal design is probably the last way you would want to design a versatile robot unless its aim is to blend in (like T-models) or improve on human characteristics (like Cylons or replicants).

Protocol robots are mostly used as tools and messengers by their owners and programmers. Be aware that they can slip by unnoticed due to their nonthreatening nature—but information is power in this line of work. Use this to your advantage and dismantle any droid that seems to pose no threat, especially the polite ones. Most can be felled with light khanjars or broken down into a unharmful collection of components and worn as a backpack.

SUDDEN BUT INEVITABLE TRAITORS

It may be startling to realize that someone to whom you've offered hospitality, firm leadership, and a second chance could turn on you, but there is actually a long list of reasons why someone close to you would conspire to stab you in the back. Jealousy, first and foremost. You're better looking and have first pick of the alien babes. You're more skilled in battle. You have your own ship. It's your face, not theirs, that decorates the reports of your collective victories.

For the weak-willed, it takes but a single worm tongue, whispering in their ear with the promise of credits or a captain-hood of their own, to convince them that *you* are the enemy. You mustn't let the overwhelming heartache and inevitability of such betrayals turn you paranoid. But when the traitors come for you in the dead of night, or sell out your position at your most vulnerable tactical moment, you must take it in stride—and then make an example of them. Doing so will increase the respect your remaining crew has for you, and discourage them from considering a path of betrayal in the future.

SPACE MADNESS

Space madness is a severe and some-times permanent condition that results from long voyages in unnat-ural environments. Experienced in weaker forms as far back as Earth's first sailors, Space madness is a sick-ness of our very DNA. It senses that we live where we do not belong, and our body fights to convince us to turn back. Or die trying.

Do you have Space madness? Likely not, or you would probably be preparing this book for dinner rather than reading it. While we're on the subject, here's how to identify the condition in yourself or a fellow crew member.

SYMPTOMS OF SPACE MADNESS

- Increasing reclusiveness
- Refusal to socialize with other members of the crew
- Sudden bursts of anger
- Loss of appetite
- Insomnia or irregular sleep patterns
- Incoherent ramblings
- Hallucinations, such as imagining fellow crew members as monsters that must be slain or bars of soap as delicious ice cream bars that must be eaten
- Homicide, suicide, or excessive showering
- Affection for clowns

If a crew member is showing most or all of these symptoms, your safest move is to kill him before he kills you first. If you are showing most or all of these symptoms, your safest move is to systematically kill everyone aboard before they grow wise and kill you first. Your next best bet is solitary confinement. If you have a room with no sharp objects or corners, you may be able to make it back to civilization in time to rehabilitate your mad Space friend or self.

On a long enough voyage, Space madness *is* inevitable, but there are some things you can do to delay its onset:

PREVENTING SPACE MADNESS

- Regular socializing
- Open, casual sex
- Exercise (such as boxing, Zero G cycling, or open, casual sex)
- Maintaining a regular schedule, despite your increasing awareness of the lack of sunrises, sunsets, or other regular intervals by which we've been evolutionarily trained
- Consuming comfort food instead of gruel as a special occasion. (Though this may result in Space paunch, and honestly which is worst? The answer is paunch.)
- Keeping a journal and rereading past entries: *Am I descending into madness?*
- Not eating soap
- Reducing length of voyages
- Increasing time planet-side between voyages
- Freezing yourself in a cryotube
- Meditating
- Punching clowns

SPACE PAUNCH

Say, there. You're looking doughy. And despite your best efforts, that beard is no replacement for an honest-to-god jawline. You're succumbing to the most dangerous Space enemy of all: Space paunch.

I know what you're thinking. It's hard enough to stay fit in normal gravity, or in an environment where you can go outside for a jog or at least try a juice cleanse. Even if it's been *months* since your last Space fight and the will to forge your body into a lethal weapon has drained slowly from your marrow ever since, these excuses are no excuse.

If you're between archnemeses, all the more reason to prepare your body for the next one. What will happen when they emerge from a black hole seeking revenge on the first creature they find? You'll either be dismissed as an unworthy adversary or swiftly eliminated. A fattening appetizer for their upcoming four-course meal of terror.

If you're bored, that's all the more reason to spend your downtime in the ship's gym. Run laps through the halls. Turn the artificial gravity on high for a life-threatening workout even as you eat and breathe. Master a new martial art. Have an ensign stitch your uniform one size smaller to *guilt* you into fitness.

You're a Space Hero! Half of what you have is your body. And your body is a reflection of the other half, which is your mind. Crunches, pull-ups, a high-protein diet, and regular mirror flexing. That's an order.

CHAPTER 17

ARCHNEMESES

An archnemesis is like a sexually transmitted Space sex disease (STSSD). With any luck, you'll only ever have one.[1] But more than likely, you'll have several throughout your life, dealing with one after another after another, over many hard-fought battles and trips to the sick bay. More likely still is that you'll have several at any given time. But there is usually one—the most itchy, the most persistent, and the most difficult to explain—that sticks out among the rest.

This is your archnemesis.

ESTABLISHING AND MAINTAINING A HEALTHY, LONG-LASTING RIVALRY

Your relationship with your archnemesis, like any meaningful relationship, will take hard work and sacrifice. Remember, until death do you part. What do you feel in your heart and loins for this villain? You must reflect on this often and work toward improving your hero-villain dynamic *every single day*.

Here are some other tips for cultivating and maintaining a lasting, healthy relationship with your archnemesis:

---

1. No Space Hero has ever made it by without any at all. Archnemeses, we mean. And also STSSDs.

- **Give each other space.** If
you're around each other
too much, you risk growing
bored with one another. Or
worse, learning to *like* one
another. Space is infinitely
large. Fuck off for a bit.

- **Establish appropriate boundaries.** You're at the tops of
each other's lists. Be confident in that. But that doesn't
mean there isn't room for other death-defying romps in
your lives. You should feel comfortable ridding the sector
of a particularly troublesome band of Space pirates, and
your archnemesis (AN) should feel equally comfortable
pillaging defenseless colonies and salvaging the shipwrecks
of peacekeeping vessels. Your AN can't antagonize you
alone—that's not a nemesis; that's a pest.

- **Make time for one another.** So you're not exclusive. You've even managed to fell several foes since the
last run-in with your AN.
Good. But you can't go *too*
long without a death-defying
encounter, or the flame will
begin to flicker. By design or
by fate, take the time to see
your arch no less than once
in every twelve adventures. If

it's been longer than that and there's still no sign of him,
your AN may be cheating. He might not even feel the
same way about you anymore![2]

---

2. *Dirk's Tip:* It's always better to know or let know early in matters of
crushing emotional disappointment.

- **Spice things up.** Any dullard can take a hostage or threaten to unload his arsenal right into his nemesis's face. A handful of such pedestrian encounters can spell boredom on the battlefield. Don't be afraid to  spice things up. Set a trap. Make a bribe. Use what matters most to your AN against him or her. Invite a disenchanted member of his or her crew to your own. Role play. The deeper the cut and the further from standard protocol, the more your heart will flutter when your arch's ship signature rings through your instruments.

WHAT ARCHNEMESIS IS RIGHT FOR YOU?

You typically don't choose your archnemesis. At first, you may not even recognize you have one. Brief scares and easily eliminated enemies crop up all the time, after all. It's only after you stop and look back—recognizing how long you've been fighting this specific person and the magnitude of your rivalry—that you'll know for sure. You've found the One.

Typically, though, your first encounter with your archnemesis will be the opposite of love at first sight. You'll know in an instant because of the way her personal philosophy mirrors, and yet distorts, your own. You'll know because of the way she just burned your family alive or took hostage the only man, woman, or alien you've ever loved. You'll know because you're not so different, you and her.

You'll know, because your arch will know…and your long, deadly tango will have begun.

POWER-HUNGRY VILLAIN

Examples: Khan Noonien Singh, Emperor Palpatine, Jabba the Hutt, General Kruge, Jean-Baptiste Emanuel Zorg, Baron Harkonnen

If you find yourself being menaced by a high-ranking member of a Space military, acting government, or crime syndicate, then congratulations! You have drawn the simple-minded straw. It's not that your villain is dumb, nor should you make the mistake of mistaking him as dumb. That would be a mistake. But your villain's motivations are more singular than most. His aim is power, and more of it. And then, after he's obtained some, more power after that.

Oh, it starts off innocently enough. He wants to be elected to parliament or captain his own ship. You can relate to that. But then he wants a bigger ship. A bigger title. Before you know it, you have a madman shooting lightning out of his fingers from the view port of a moon-sized battle station with enough firepower to destroy a planet with the push of a button.

Why should anyone need to destroy an entire planet all at once? How impractical. How does it maneuver? Where did they get all the metal to build it? Wouldn't it be pummeled by the debris of its very first victim, or be susceptible to the myriad design flaws that were likely implemented in its very hasty construction?

In a word: yes. And it's up to you to stop him. Your power-hungry villain's thirst for power can never be quenched. Parch him in the face.

CORRUPTED-BY-EVIL VILLAIN

Examples: Darth Vader, Darth Revan, most Sith Lords, Saren Arterius, Captain Bialar Crais, the Illusive Man

It may come as a bit of a shock to a budding icon of wholesomeness such as yourself, but the forces of evil can be quite inviting. Evil promises high rewards at a low cost. Villains corrupted by evil are, by definition, individuals who were good once, or *could have been good*, if not for a tragic intervention beyond their control: a traumatic childhood event or the loss of a loved one. These tragic souls begin to work their way through the five steps of grief and cash in all their chips at "bargaining." If only they had *power*...they could make all things right again.

Villains corrupted by evil are often difficult to identify while they are arching you. As you look down upon their villainy from your gilded heroic perch, there would seem to be nothing redeemable about your murderous foes. But if you were to get to know them, you may find that there was some good in them after all.

For most arch-caliber villains, there is no *true* redemption for their evil deeds outside of death. On the plus side, you can help them achieve this redemption! Good on you.

VENGEFUL VILLAIN

Examples: Khan Noonien Singh, Captain Bialar Crais, Nero, General Thade

Chances are, even you will see your fair share of tragedy. Your home world ripped asunder by war. Your family burned to death and left to crisp in the setting suns. You'll lose countless crew members and possibly even lovers that you thought you really might settle down with this time. Imagine, however, if the common afflictions of Space life were to strike you and you *weren't* an awesome, good-looking, quick-witted Space Hero. You might turn a bit mad. You might even turn…to revenge.

Revenge villains operate under the delusional notion that hurting those who have hurt them will help heal their wounds. The great tragedy of their lives is that they often don't realize the futility of their mission until it is too late. They kill those who killed their loved ones and feel nothing. They use unregulated cloning technology to bring their loved ones back from the dead and are cursed with soulless living reminders of all they've lost.

Unable to reclaim what has been taken from them, revenge is often a stepping-stone to greater forms of villainy. And it is a slippery one.

FIERCELY PRINCIPLED VILLAIN

 Examples: HAL 9000, Colonel Miles Quaritch, Yautja (Predators), the Operative, the Illusive Man

You're a Space Hero of principle. No unwarranted violence against those who have not yet violenced against you. No comrade left behind. No open beverages in the cockpit. Your principles, some born naturally within you, others learned living through the trials and tribulations of Space life, are what make you *you*. They inform your decision making, your unassailable reputation, your infallible judgment of character, and your fashion sense. Your principles bend, but do not break.

Now imagine someone as fiercely principled as you, as committed to his or her principles as you…but with a different set of principles entirely. This villain believes the right-sized purse is worth the death of any man, wears sleeveless fur coats in all weather, and has toilets that flush counter-clockwise.

Often, principles are more dangerous than thirst for power or revenge, because a fiercely principled villain may not realize she's the villain at all. Be ready to teach her how wrong she is, possibly through brain damage.

RELENTLESS OR IRREDEEMABLY EVIL VILLAIN

Examples: The Borg, the Brood, Arachnids, Xenomorphs, Darth Nihilus, HAL 9000

Some evil cannot be explained away. Some evil simply *is*.

The hippier among you may attempt to explain away innate, irredeemable evil as a by-product of nature. If a species emerges, evolves, and thrives, what fault is it of theirs if that success is based on compassionless xenocide?

Well, hippie, when half of your crew has been slaughtered, and another third has been assimilated or turned into biological incubators for the brood of whatever monster now hunts you, there won't be time to wax philosophic about the true nature of "evil." If it wants you dead, and it can't be reasoned with, it qualifies.

The terrifying by-product of irredeemably evil villains, as if it could get any worse, is that they are proportionately relentless. They come in waves, groups taking the place of individuals, the still-living taking the place of the dead. They come until you put a stop to them.

ENDING THINGS

Even the most thematically meaningful archnemesis relationships must someday come to an end. Eventually, one of you must die at the hand of the other.[3] And though you hate your arch with every fiber of your being, it will not be easy. It may be quite painful. You've put in so many good years, after all. You may have even, in your AN's final moments, removed the dark mask of his or her true nature, revealing redemption—maybe even good—underneath.

Do not let such redemption and resulting empathy sour your victory. You've earned this. Without you, your AN may never have had a chance at the last-second redemption of his mangled, black soul, and may have afflicted the Universe with a great many more evils if you had not been there to stop him.

There may never be another archnemesis for you. You never forget your first, after all…But his legacy will help to define your own. Without evil, there could be no good, so perhaps, in that way, it is good to be evil sometimes.

 THE SPACE HERO'S WORKBOOK:

What Kind of Villain Is Your Archnemesis?

_____

What Is the Name of Your Archnemesis?
(The Weirder, The Better!)

_____

---

3. And it had better be him by yours, or you're not very good at this.

CHAPTER 18

ALWAYS SHOOT FIRST: A BEGINNER'S GUIDE TO COMBAT

S pace: the Final Frontier. Space: also a big, open hellhole where there are virtually no laws, and people take what they want from you, leave you to die, and laugh while they do it.

If you want to survive long enough to do any good, there's one lesson you need to learn:

Always. Shoot. First.

That doesn't mean "blast people in their faces at the slightest provocation," but it does mean that you've got something to learn from the man considered perhaps the wiliest of all Space Heroes, the bounty hunter turned smuggler turned carbon Popsicle turned general turned nerf mogul, Han Solo.[1] In one of his most famous acts, Solo reportedly gunned down a man in a bar, a bounty hunter who was looking to collect a reward in exchange for Solo's corpse. Held at gunpoint, Solo carefully distracted the bounty hunter with talk and platitudes, while slipping his own gun out of its holster beneath the table, and then calmly took the bounty hunter out.

---

1. That's "nerf," the shaggy, bovine-esque Alderaanian quadruped often kept in herds and prized for its delicious steak, although Solo was known to have his fingers in a few foam dart toy-related merchandising ventures as well.

Some accounts state that Solo only returned fire after the bounty hunter fired on him and somehow missed at point-blank range. These fabrications, created by historical revisionists trying to cast Solo in something of a more favorable light, are dead wrong. Solo's willingness to open fire on a bounty hunter holding him at gunpoint doesn't make him *less* of a hero. It makes him *more* of one, and one who ought to be emulated.

In this chapter, we'll discuss the ins and outs of combat, from the rules of engagement to the proper management of weapons. You can't be an effective Space Hero if you're Space dead. So ask yourself: What would Han Solo do?

HONORABLE FIGHTING AND "HAN-ORABLE" FIGHTING

A Space Hero has a reputation to uphold. That reputation is one of protecting the innocent, choosing mercy over revenge, and opting for peaceful resolutions over ass kickings unless absolutely forced.

But a Space Hero also has a crew and a self to keep alive. Doing so will frequently lead to choices that threaten the sanctity of his or her reputation. What you need is more than a reputation: you need a philosophy. Let's chisel one out.

SHOOT FIRST

From the legendary story of how he outplayed a dim-witted bounty hunter to the lesser-told tale of when he simply starting unloading blaster bolts on a high-ranking government official the second Solo recognized him at a Bespin state dinner,[2] when faced

---

2. To be fair, that guy was an asshole. And anyway that guy was fine (according to the stories).

with imminent danger, good ol' Han didn't blink—he just did what needed to be done, "second-degree murder charges" and "bar etiquette" be damned.[3]

If the choice is between shooting first, potentially causing a diplomatic maelstrom in the press, and shooting second, potentially being eulogized at length in the press, you don't want to be shooting second.

SHOOTING SECOND—AND THIRD, ALSO

Well, okay, you do want to shoot second. And third. And as many other numbers in sequence as are necessary to neutralize the threat. Remember: a warning shot counts as a miss.

COMBAT: A BEGINNER'S GUIDE

You have a natural talent for Space Heroics (you wouldn't be reading this book if you didn't) and will acquire the skills required to maximize that talent over time. But even if you're just starting out, the basics of weaponry and their associated forms of combat will earn you that time—and the respect of your growing list of enemies.

HAND TO HAND

A good Space Hero doesn't need a weapon to be an effective defender of the innocent and mangler of the wicked. He or she doesn't need steel, light amplification by stimulated emission of radiation, or sonic electronic ball-breakers. A Space Hero needs only his or her wits and a few choice combat tactics.

---

3. Although one must note that "Good Guy" Solo paid the barkeep for the mess he caused. Space Heroes think about the little people.

BALL BOTH FISTS TOGETHER AND SWING THEM LIKE A CLUB

A no-brainer, given that if one fist is an effective weapon, *both* fists would be an *even more effective weapon*. Whenever possible, a dignified Space Hero doesn't just throw punches like some sort of low-rent television action hero. Throwing punches can put you off balance, send you careening into bulkheads, and even badly injure your good pointing hand. How are you going to constantly yell at minions to make it so if you can't accurately point at the "so" they should be making? Instead, whale on baddies as if you're holding an invisible baseball bat by planting your feet, swiveling from the waist, and using the power of your angry face. Put your spine into it.

COMBAT ROLL

Sure, you could run three steps to get into position to stab or shoot your enemies, but why run when you could *roll*? It's a proven fact that a forward somersault from a standing or running position is a perfect evasion of literally any form of attack—from a rancor-bone spear to a light-speed phaser blast. Any time you think you might want to cross open ground on foot, you should consider a roll instead. It's only moderately disorienting!

NERVE PINCH

Why wallop your enemy with the two-handed ham-fist attack when you could stealthily take away his ability to stand under his own power? With the Nerve Pinch, now you can! An ancient Vulcan martial art akin to Earth's "purple nurple" or "atomic wedgie," the Nerve Pinch requires the user to simply grab the right spot[4] to inflict blinding, searing pain on the subclavian artery

---

4. *Dirk's Tip:* May also require a heaping dose of Sneaking Up From Behind.

and a certain nerve ending, and *boom*! Unconsciousness without a fight. It's a technique that belongs in every Space Hero's arsenal.[5]

PERSONAL SHIELDS

Depending on whether you're the kind of Space Hero who goes around in sealed armor all the time, or would prefer to hang out in your spandex onesie uniform, a personal shield might be for you. Those heroes constantly engaged in intense fights will find these shields useful, since kinetic barriers can stop most bullets and energy beams, and blades and swords take special training to penetrate them.[6] You'll need some kind of power source to carry around with you to generate the shield, though, which is still considered a fashion faux pas in most sectors.

SIR ISAAC NEWTON, ZERO G PRIZEFIGHTER

Sir Isaac Newton of the seventeenth century was a premature baby who grew up to be a frail math nerd who struggled with the ladies…But he would box your ears and upend your powdered wig in Space, by Jove.

That's because Sir Isaac Newton understood the laws of motion. In fact, he

---

5. Note: It may require some kind of latent telepathic ability to pull off, as it doesn't seem to work for everyone.

6. Know your shield. Some technologies of personal energy shield, such as those common to the Padishah Empire, can explode when they come into contact with certain laser weapons, which tends to defeat the purpose of the shield.

came up with them. And in a frictionless environment (like, say, Space), Newton's laws of motion are at their most applicable. Raw size, brute strength, and pickup artistry won't help you here, burgeoning Space Hero. In a zero-gravity environment, they might even hurt.

Using each of the three laws, let's review how a genius poindexter would present your own backside to you in a zero-gravity fight.

First Law:

"Every body persists in its state of being at rest or of moving uniformly straight forward, except insofar as it is compelled to change its state by force impressed."

The first part of the first law isn't of much concern to us in the area of Zero G fisticuffs, but to state the obvious: with no outside forces at work, and no pushing off something of your own volition, you're not going anywhere.

The second part of the first law is much scarier in the context of theoretically infinite space. Suppose you're on one side of your spaceship, and you kick off the wall to get to the other side of the spaceship. The force of your legs into the wall and the wall pushing back (more on that in the second law) sends your body through the frictionless environment at a single speed and in a direct line. This combination (speed plus trajectory) is known as velocity.

Being able to predict your speed and destination with minimal force is rather handy. The supper alarm's blaring, for instance, and you're a quick kick from being first in line for Space gruel. But suppose for a moment that opposite wall you're aiming for isn't there. Without anything in your path to stop you or slow you down, the slightest push would send you hurtling through Space (or at least spinning very slowly through Space). Forever.

How Sir Isaac Newton Uses the First Law of Motion to Present You Your Backside

Sir Isaac Newton knows that, as easy as it is for you to hurl yourself at him bodily, it's just as easy for him to dodge out of the way and watch as you impale yourself on a ladder. Sir Isaac Newton also knows that ship walls, inside or out, are the only barrier between you and an eternal Space burial. One careful dodge or removal of your landing zone and you become a ridiculous cartwheeling speck, and he becomes Señor Newton, Space torero.

SECOND LAW:

"The alteration of motion is ever proportional to the motive force impress'd; and is made in the direction of the right line in which that force is impressed."

What the second law is referring to, if you're up to speed on your Latin (translations), is acceleration. The greater the force acting on an object, the proportionally greater its acceleration will be.

As with all the laws of motion, there's some overlap here. Recall the example above in which you kicked off the hull of your spacecraft to propel to the mess before your hardworking crew? The harder you kick off the wall (the greater the force), the *faster* you'll go flying through the void.

How Sir Isaac Newton Uses the Second Law of Motion to Lay Low Your Malodorous Carcass

Mathematically speaking, the force being exerted upon any object can be measured by its mass times its acceleration. This means a superbrain like Sir Isaac Newton needs only a few quick mental calculations to figure out how easy it would be to give you a concussion. Even more embarrassingly, he only needs to see you hurl yourself toward him once to figure out exactly how much you weigh, and therefore how far you've slipped on your diet and exercise regimen since you were last planet-side.

Sir Isaac Newton knows it's not the size of the wallop—it's how you use it.

THIRD LAW:

"To every action there is always an equal and opposite reaction: or the forces of two bodies on each other are always equal and are directed in opposite directions."

The last law of motion is the most important in matters of physical altercations in gravity-free and atmosphere-free environments. Like the second, the third works in tandem with the other laws. Recall again the example in the first law, in which you kicked off the wall to propel yourself to the front of the gruel line. It says explicitly that you kicked the wall, and the wall *pushed back*. If it didn't, the energy of your kick would be absorbed into the wall and you wouldn't go anywhere.

How Sir Isaac Newton Uses the Third Law of Motion to Hand You Your Ass

Those muscles of which you're so proud? Those mok'bara moves you've tirelessly honed? Useless against Long-Nosed Ike. For he knows that in this Zero G arena of death, each blow that you land on him, you will also land against yourself. The force with which you bludgeon him will only serve to send you flying backward with *equal* force in the *opposite* direction, giving up your position and hurtling you across the arena.

This applies to standard projectile weaponry as well. Maybe you're Space Hero enough to handle the kickback of a sawed-off 14-gauge shotgun in normal gravity, but that sucker will shoot you back like a human-sized bullet every time you pull the trigger in microgravity.

Now that you've been knocked down a peg by a skinny white mathematician with bad hair, you should have a preliminary grasp of Zero G combat and the size of your puny ego.

LASER GUNS

Lasers—the perfect Space weapons. Unlike standard projectile weapons, lasers use the physics of light and energy to make them deadly over vast distances and at great speeds, not bothered by pesky nonsense like friction or explosive powders to fire them.

Your standard laser beam works by stimulating photons and sending them out at a specific wavelength and frequency—which basically means that the waves of the beam don't interfere with one other, allowing them to send energy efficiently from one place to another. What *that* means is that you point your laser gun at someone, and not too long later, they'll need a new silver Space onesie or organs or both.

Knowing how your particular laser works is very important.

What if you need to repair it in the field while under fire from Tavlek mercenaries who have kidnapped your deposed dominar traveling companion? What if you find yourself in a batarian prison and have to fashion something to cut the bars out of ancient PlayStation parts? What if your laser suddenly starts making weird noises and you're worried it might overload in your hip holster and blow off your external reproductive bits? Knowing your weaponry as you would your own appendages can save your life—and end the enemy's.

WHAT KIND OF LASER IS RIGHT FOR YOU?

There are numerous directed-energy weapon options available to the standard Space Hero. You'll need to choose the right one for your particular needs.

Standard Laser

As described above, your usual laser creates a coherent, high-energy laser beam through the release of photons. That's done by stimulating a substance called a "lasing medium," usually a crystal or gas, with radiation or electricity. The atoms in the medium are excited and their electrons gain more energy. To reduce their energy and attempt to maintain equilibrium, the electrons give off photons.

Those photons get bounced off mirrors to excite more atoms and release more photons, then get passed through a lens, which further contracts their wavelengths and creates parallel waves that don't interfere with one another, and the beam is emitted. With enough energy and at the right wavelengths, lasers can become dangerous and deadly—it's all about concentration of impact and energy conduction. In the simplest terms, it's a superconcentrated flashlight that melts people and their stuff.

Phaser

Starfleet-style phasers—or PHASed Energy Rectification—fire beams of "rapid nadions," a form of subatomic particle with which it sucks to be intersected. Phasers use superconducted plasma to divide atoms, sending protonic charges in a concentrated beam. It's the Swiss Space Army Knife of laser-beam sidearms, both a powerful weapon and a multifaceted tool that can be used to cut through bulkheads, heat up rocks, and vaporize biological and mechanical life.

Blaster

The First Galactic Empire and its preceding Republic altered their laser technology into plasma-based, wallop-packing "blasters." Though similar to lasers, blasters turn small amounts of high-energy gas into concentrated particle beams, then combine it with high-intensity laser energy to create a blaster "bolt." The ratio of superhot plasma to superhot laser can determine the power of the bolts. Typically they are powerful enough to blow holes in walls as well as plastic armor as well as the folks inside. Blasters require both gas and a power cell for "ammo."

SPACE SWORDS: AN ELEGANT WEAPON FOR A MORE DISMEMBERED AGE

Even in an age in which better technology is everywhere you look, everybody who's anybody among Space Heroes (and most Space villains) carries a sword. Sure, a projectile weapon is safer for you and more effective on your enemies, but anybody can fire a gun. Only a complete idiot or someone who's so deadly they straight-up don't give a shit uses a sword.

And that, Space Hero, matches your description.

SPACE SWORD VARIETIES

Not every hunk of sharpened, folded steel is acceptable for a captain to carry. This is *Space*, after all—there's decorum to uphold, a certain standard of awesome. You're not going to want to carry anything less than a qualta blade into battle with you. More than likely, you're going to need something of the vibroblade variety just to stand up to your various laser swords and plasma blades. Don't know what any of those words mean? Don't worry, the only question you need to ask is, "Will they cut Space stuff?"

Yes. Yes they will.

D'k tahgs, Crysknives, Qualtas, Lirpas, and Other Ceremonial Blades

Whether you're dealing with a rage-boner in a Vulcan koon-ut-kal-if-fee against your best friend, or just knifing fools in bar fights with the d'k tahg you took off a tipsy Klingon, you can do worse than any of the billions of ceremonial blades floating around the 'Verse.

Vibroblades

Somewhere in a galaxy far, far away, someone thought to combine the faux-samurai sword available in your local Space mall and the electric carving knife your grandpa had to use at Thanksgiving because of his couch-induced muscle atrophy—and the result was the vibroblade. Basically, it's a sword that vibrates incredibly fast, so that instead of just cutting stuff, it cuts stuff *more*.

Throw in your standard cortosis-weave, if you happen to have some cortosis ore around, and you can put your vibroblade up against the standard laser swords, light scimitars, and energy blades you might come across.

ENERGY BLADES AND SABERS OF LIGHT

The coolest of weapons available to Space Heroes, energy blades are literally blades *made of energy*. They not only are easily the coolest technology of all technologies, but they also carry that mystical "whoa, that guy can manipulate light and use it to hack off your arm because that's what a badass he is" factor.

Your standard sword composed of electromagnetic radiation is lightweight[7] and carries incredible slicing power, since its blade is made up of particles that are small enough to shear through its prey at the subatomic level. "But wait," you ask yourself, "shouldn't light beams continue in a straight line and not magically stop at a convenient sword-length height?" *Yes, of course they should*, is the answer—except that beam swords are smarter than you.

How Light Rapiers Work

It's a little-known fact that a sword made of light is not actually a static laser beam, but energy converted into plasma through the sword's various internal components. The machinery in the light rapier's hilt creates a circuit, with energy flowing through the beam as it would through electronic circuitry and bouncing back into the power cell, thanks to the influence of a force field.[8] Crystals focus the beam and spray out enough particles to make it all stick together in a nightmare of searing-hot minirave.

---

7. Yes, of course that pun was intended.
8. Not to be confused with the Force field.

Energy Blade Usage

The first thing you need to know about energy blades and laser swords is that, cool as they look and simple as they seem to be to use, picking one up and waving it around like an idiot is a great way to wind up a limbless torso shrieking in pain as your enemy stands over you, laughing (or more likely, lecturing you about your bad choices). Light spathas and their ilk are known to create intense gyroscopic forces, making them resistant to motion. Couple that with the fact that plasma doesn't weigh anything, and it becomes very easy to accidentally slice your padawan in half right before you try to take down that super villain.

On Blocking

You've probably asked yourself, "Hey, how come a blade made of light (or plasma, since you're learning) doesn't just flash right through everything, including other blades made of light (or plasma)?" That's a very good question answered by Science. Specifically, light katanas are made possible by the containment fields that keep the plasma in your basic sword shape. Those fields block the transmission of plasma through them, but not other things (hence cutting), so when two light jians come together, the fields block their plasma from continuing through each other (hence awesome sound effects).

When Is It Okay to Space Sword Fight?

There are very specific situations when drawing your sword is the thing to do, since bringing a knife to a laser fight is a good way to get various essential body parts disintegrated. Those situations are when your adversary *also* brings a sword and is committed to a badass showdown of badass proportions.

The short answer, then, is that it's *always* okay to bring a sword to a gunfight, turn a gunfight *into* a sword fight, and, generally, be looking to sword fight, provided you've got someone to sword fight with. But, you know, maybe have a friend standing by with your blaster just in case.

———

Feel the blood of battle coursing to that one weird vein in your forehead? Are the bodies of your enemies laid around you, limbs sprawled to spell out the phrase, "WE'RE SORRY"? Good. Your combat reputation shall soon precede you.

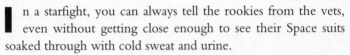
In a starfight, you can always tell the rookies from the vets, even without getting close enough to see their Space suits soaked through with cold sweat and urine.

The rooks are easily disoriented. They operate under the same assumptions as planet-side fighter pilots. There's ground beneath them (perhaps a nearby planet), the sky above them (the distant stars), and the battlefield (more or less an even plane in between).

As soon as the Space battle begins, these comforting orientations shatter, and the cold sweats and urinating begin.

The vets know that there is no ground below and sky above in Space. Should they be lucky enough to be near one, sure, a point of reference can be handy, especially when a hasty retreat is required. But the vet also knows that abandoning traditional notions of up, down, forward, and back is not only necessary to survive the fight—but an advantage.

WHICH WAY IS UP?

If you answered "That way," with or without a pointy finger, you are *wrong*. A trick question to start us off, rook. In Space combat, there is no "up." The sooner you'll learn that, the sooner you'll draw comparisons to war hero and bug squasher Ender Wiggin. This is something you should covet.

It is common for Spacefarers and Space fighters to move through the air lock and become suddenly overwhelmed with vertigo. The ship is at their back, the walls pass beside them, and they feel as though they're moving *forward*, all systems nominal—*until* the bay doors slide open, forward becomes down, and their lunch becomes moist potpourri on the inside of their helmet.

For a special few, this experience will pass. For salty vets, the cascading formations and kaleidoscoping explosions *are* their earth and sky, both.

Once you've overcome the puking phase, here are some ways the disorienting nature of Space can help you kill countless thousands and their unborn children.

- Abandoning concepts of up and down turns a two-dimensional field of battle into a three-dimensional control scheme. Why fly directly into the face of the enemy, their weaponry locked onto you all the way, when you could come at them from the side, from behind, from overhead? Your angles of attack are myriad.

- Unless the engineers back at Central Command have come up with something truly revolutionary,[1] your ship has some sides that are broader than others. Bigger, longer, broader sides make for bigger, longer, broader targets. No matter your direction, make your ship the smallest target possible in relation to the enemy, and come at them from an angle that gives you the biggest chance to land a killing blow (although it's worth

---

1. Something they haven't done since the invention of the helmet straw, in this Space Hero's opinion.

noting that you'd better have a bunch of guns on the small side in question).

- In a spherical battlefield, your routes of retreat are infinite. The shortest distance between two points will always be a straight line, but a zigzaggy loop-de-loop may be the shortest distance that doesn't involve your emulsification.

MANEUVERS YOU SHOULD MASTER

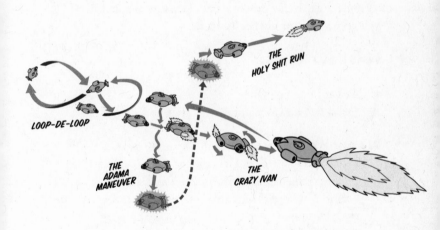

LOOP-DE-LOOP

THE HOLY SHIT RUN

THE ADAMA MANEUVER

THE CRAZY IVAN

THE ELEMENT OF SURPRISE

Doesn't exist in Space. At least, not without a little creativity and high doses of Hollywood-grade luck.

In most ships, instrumentation is just too sophisticated for a sneak attack, since there's often nothing but Space between you and the thing you mean to attack as you approach it. Your theoretical enemies will see you coming full days away. Even the naked eye is a surprisingly effective tool for detecting bogeys. The first burst of thrust of an engine or glint of sunlight off a view port, and you're made.

Here are some notable exceptions, which may be used alone or in tandem.

GO DARK, A.K.A. THE WASHOUT

Set a course, kill the lights, drop the engines. Your inertia will carry you the rest of the way, and your heat signature will decrease as you approach your intended target.

✚ WASHOUT PROS:

- Cover of darkness
- Decreased heat signature

⊖ WASHOUT KHANS:

- Highly mathematical maneuver—you'll need tons of calculations to make sure you intercept your target at the right location
- Gradually frozen to death
- Decreased visibility
- Decreased maneuverability (straight shot or one spot 'til go time)

HIDE UNDER THEIR NOSES, A.K.A. FALCON NEST

This is more commonly used as an escape tactic—park your bird in the instrumentation dead zone of a larger ship, such as against its hull, until they're sure they've lost you. But if you can hide under their noses long enough for them to think you're gone, *then* reappear again, that might just make the opposing Number 2 drop his namesake into his standard issues.

⊕ FALCON NEST PROS:

- Avoids proverbial foot race
- Avoids actual firefight
- Makes enemy look dumb, which is hilarious

⊖ FALCON NEST KHANS:

- Preeetty good chance they'll find you
- Difficult to time your reappearance and escape
- You're going wherever they're going

DESCEND FROM THE RAFTERS, A.K.A. SCARFACE

Obstacles are few and far between in Space, but if you can lure your enemy past a lunar surface or, better yet, an asteroid field, you can wait until they float unwittingly past and get the drop on those alien or rebel or Imperial scumbags. Works particularly well with the Washout.

⊕ SCARFACE PROS:

- Ensures a direct hit[2]
- Asteroids provide cover
- The enemy may conveniently explode against Space rock themselves

---

2. Unless you've got the aim of a storm trooper.

SCARFACE KHANS:

- They may not follow you into that asteroid field
- You also need to go into that asteroid field
- You might be waiting there a looonng time

STAY STUNNING! WITH DIRK PARSEC

Laser Pollution: Help Save the Icy Black Void

The probability of a wayward laser ever striking something once seemed so astronomically low that laser pollution was thought to merely be the latest hippie boo-hoo. But after hundreds of wars and thousands of chases, laser pollution has grown into a real concern. What we hadn't considered as we hurled thousands of pastel-colored photons into the black, icy void over the course of hundreds of years is that *lasers keep going*. Forever. Or, until they hit something. Whichever comes first.

Take Wedge McAllister III. Private first class on the side of the Argonauts in the Great Battle at 47-Tentacled Rock in 2583. At only eighteen years of age, Private McAllister fired many lasers into the flesh of the Centurions that day. But he fired many more into the ether. As we are wont to do, we don't consider the lasers that miss during a fight. If it hasn't hurt an enemy or hurt a friend, it's of no consequence.

But somewhere out there...Private McAllister's misfires were still soaring at the speed of light.

Wedge went on to become a purveyor of fine kitchenware, mix DNA with a Shemalien, and raise four quasi-proud children. It wasn't until Wedge was sixty-eight years old, fifty years after 47-Tentacled Rock, when Wedge made a house call to the Duke of Omega 19 about some porcelain teacups, that his story came full circle.

There he was, bringing his Martian commercial freighter into port, when a wayward laser blasted through his shield-less view port and struck poor Wedge in the heart. He was dead in an instant.

The duke was of course held and questioned until Omega CSI picked up a communiquè from their top consulting physicists that cleared him. The physicists had traced the route of the laser that felled Wedge, based on the angle and surrounding heavenly bodies and determined something astounding:

The laser beam had originated at 47-Tentacled Rock, fifty years prior.

Somehow, the laser had shot through Space, bounced off radioactive waves, curved around suns, and found its way to the exact spot where Wedge stood aboard his ship, fifty years after he had fired his weapon for the Argonauts as a private first class.

Against all odds, felled by his own weapon.*

In death, Wedge McAllister III has become the face of laser-pollution awareness. Wayward lasers have struck colonies, pierced mining ships, redirected meteors, and pointed us back toward our history of senseless overfiring. But it's the laser shots we *can't* account for that are the scary ones. The laser shots that still haven't found their target.

Is this who we are? Is this the legacy we want to leave to our future Space Hero babies? A future where Space travel is unsafe, even for the simple slinger of fine teacups, for fear of being struck down by a laser meant for someone else?

The solution begins with us, Space Hero. And that solution is: don't miss. Ever.

*Or close enough. Let's not split atoms over a coincidence of this magnitude.

INNER COCKPIT COMBAT: COMBATING G-FORCES

The G-Forces are a fearsome gang of rapping pirates. In general, you should not be doing battle with them inside your cockpit. Should you engage them in a rap battle, fear not—their rhymes are hella weak. For purposes of this section, which is confusing and mathematical enough as is, forget they were even mentioned.

The g-forces you are much more likely to encounter in your cockpit are a measurement of acceleration felt as weight. For maximum insult to bone-crushing injury, the name "g-force" is a misnomer in two parts:

- The g in g-force stands for "gravity," but a g-force by definition *does not include gravity*. In fact it is a vector quantity (directed in a single direction) of the net forces acting

on an object *other* than gravity. But standard Earth gravity (a directional force of roughly 9.8 meters per second
squared) is the unit of measurement used to calculate an
object's vectorized acceleration.

◉ A g-force is not a force at all. G-force is the measurement
of the acceleration of all the forces acting on an object
(in this case, you), in a particular direction, but is not a
force itself.

G-forces, depending on their magnitude, can be that light
press on the chest as you bank around ancient trees on your
speeder bike or the feeling of weightlessness in their absence
(like, say, when you're orbiting a planet or have your ship's
artificial gravity generators deactivated). But they can also cause
tunnel vision, light-headedness, and unconsciousness due to a
lack of blood flow to the brain and, at extreme levels, make your
eyes burst, turn your bones into gelatin, or kill you outright.

The reason any g-force above 1 (net acceleration at 9.8
meters per second squared) *doesn't* kill you outright is that your
body is malleable—to a point. Slap your second-in-command to
knock some sense into him or jar loose a brain parasite, and your
open palm may impact his too-stubborn-to-know-what's-good-
for-him cheek with 100 g's.

But his face will take the impact and return to its normal
shape without doing permanent damage (while hopefully receiving the intended message). It's when the g-forces are too strong,
such that they push your cellular structure beyond its range of
malleability, or when g-forces act upon your body for too long,
that they can be even more dangerous than a slap from your
first officer.

As long as you're moving at a constant velocity, you

will not be impacted by g-forces. This is why on a planet's surface, which is rotating at a blindingly fast speed, you do not perpetually feel as though you're going to upchuck. It's why, even at warp speeds, you don't turn into a puddle of gooey biohazard. It's the *acceleration* to warp speeds that can kill you horribly.

This becomes a particularly tough nut to crack in the field of battle, as the chaos of a tactical firefight by definition includes jarring stops, starts, retreats, and races. The constant accelerations and decelerations are enough to make the helmets of even the most hardened gunslingers goldfish bowls of stomach spew.

So how do you deal? Puke first, ask questions later? Could do. Turn every Space battle into a war of iron stomachs, like a gaggle of middle schoolers at the county fair? Your odds at a win streak are weaker than a credit flip. And as we know how much you hate to be told the odds, here are a couple alternatives.

- **Maintain a constant speed.** As often as possible. Unless you're caged in by an opposing army, a feature of Space is that it's boundless. You can keep your fighter at a single speed and take all the Space you need to dodge, loop-de-loop, and carom in and out of battle. This reduces your encounters with g-forces to fighter launch, hasty retreats, and times of great need.

- **Anti-gees.** In a word: drugs. Drugs that keep your stomach calm, your blood coursing, and your mind in the game. The hangovers make your hangar-deck wine-pong aftermath look like grade school, and some "scientific double-blind studies" show that every injection of military-grade chemical stew can lop up to a year off your life.[3] But if you're going to take a laser up the tailpipe without 'em, what's the difference?[4]

● **Pick your spots.** With enough training, a sharp eye, and a quick trigger finger, you don't need to be the fastest smear on the battlefield to be an ace pilot. Well-timed accelerations can buy you the millisecond you need to blast a toaster out of the air and breathe to fight another day. Piloting more won't necessarily make your body more resistant to g-forces, but you will learn how much you can take—and when a little vomit and a broken blood vessel are worth trading for your life.

Now that you're safely back aboard your ship, and you've successfully avoided a stomach backflip, *brush up on your rhymes and don't waste no time. The G-Forces pirate rappers gonna turn this scene to a crime.*

SHIP WEAPONRY

You can fly, but can you shoot? If not, many of these will do the killing for you. Merely equip the enemy-eliminating weapon of your choice, and press the appropriately colored button in your enemy's general direction.

LASERS, PHASERS, AND ALL KINDS OF OTHER -ASERS

The best ship-mounted weapons tend to be of the directed-energy-beam variety, for all the same reasons you want one on your hip, draw-ready. Lasers can travel great distances and deliver

---

3. How are we supposed to trust scientists who are blind in both eyes?
4. PSA: Don't do drugs.

high-powered energy at the speed of light, which makes them ideal for firing at other ships. Whether they be turbolaser turrets—in effect, just scaled-up blasters—or phaser banks—which draw on the power of ships' warp cores to increase their strength—beam weapons are a one-stop solution for all your pew-pewing needs.

CONVENTIONAL PROJECTILE WEAPONS

If lasers aren't your thing, consider bullets and artillery shells. While not quite as showy as charged energy and plasma beams, bullets can be just as effective, particularly for militaries that lack energy-shield technology. They also gain the added effectiveness afforded them by the lack of friction through a vacuum. And though conventional projectiles require air to fire (the vacuum snuffs out combustions and explosions before they begin), your ship's ability to sustain internal atmospheres sidesteps the issue.

TORPEDOES AND MISSILES

When your target isn't shielded, consider guided missiles and other explosive weaponry. Because of a need to cut back on construction costs and the energy and fuel required to push ships around the galaxy, spacecraft hulls generally are as low-mass as possible. Once their deflector, reflector, refractor, and plasma shields are down, gather up the impressionable children for a fireworks display of the silent but deadly variety. In the event of additional shields, or piercing the hull requires something with a little more punch, proton and photon torpedoes each provide their own brand of hurt to put on your foes (see below).

NUKES

When in doubt, go with the loudest, clearest message in your arsenal. Nukes have two main destructive capabilities: they deliver a lot of destructive force in a small area (which means ships need to be specially armored to withstand their blasts), and they emit high degrees of electromagnetic radiation (in addition to particulate radiation that can ravage biology if a crew isn't sufficiently shielded), which can disrupt electronic systems.

PROTON TORPEDOES

Like nukes, but more protony.

PHOTON TORPEDOES

Packed with the devastating power of…light, photon torpedoes are actually shielded torpedoes that pack a matter-antimatter mix. When this mix is combined within the warhead, the torpedoes release a high yield of explosive power that can ravage a ship. Their shielding can even allow them to penetrate starship armor. The photon name, presumably, comes from the annihilation of matter and antimatter producing only one kind of particle—photons—with the rest of the energy produced being kinetic energy, which makes for the deathy, shreddy, boomy part.

GRAVIMETRIC TORPEDOES

A favorite weapon of the Borg, gravimetric torpedoes use gravitons in place of matter-antimatter warheads. When the warhead impacts a ship, the warhead's high-powered release of gravitons can create varying gravitational fields within the target, basically using concentrated gravitational forces to tear the ship apart. File under "Things that suck to have happen to you."

SHIP DEFENSES

As you shoot, so too will you be shot at. Begin rhythmic chanting: it's time to play some DE-FENSE.

DEFLECTOR SCREENS AND ENERGY SHIELDS

Your basic energy-based defense mechanism can disperse energy impacts at the cost of its strength and the stability of components that maintain it. Most shields can withstand contact with both energy and matter for a finite amount of time by absorbing and deflecting. They'll save you from the first blast or three, but you're on the clock, Captain. Hope you have some life-saving magic up your sleeve.

STARFLEET SHIELDS

Powered by a ship's warp core, Starfleet shields are generated energy fields created by dense fields of gravitons projected in bubbles around ships—either at some distance or just a few meters off the hull. Shields are usually created by a series of emitters set in the hull of a given ship and are projected at given "frequencies" that allow them to permit weapons fire from the ship in question to pass through the shield while deflecting incoming fire.

MASS EFFECT SHIELDS

The use of mass effect drives that run on Element Zero allows users to create energy fields that either drastically reduce the mass of an object or drastically increase it, and otherwise affect the movements of objects. Ships and personnel use mass effect field "kinetic barriers," repulsive mass effect fields, to turn away incoming projectile weapons fire. Kinetic barriers are designed specifically to deflect fast-moving projectiles, so they have less effect on slower or stationary objects, like docking ships.

HOLTZMAN EFFECT SHIELDS

The Holtzman effect allows the Spacing Guild ships of the Padishah Empire to achieve faster-than-light travel, but the phenomenon also is responsible for all manner of energy-shield technology, including personal body shields. Those shields are often upscaled to cover entire buildings and vehicles, and can be used aboard starships, although it's generally a bad idea to do so. Like personal shields, Holtzman effect shields are great for deflecting projectile weaponry and matter, especially when it's moving quickly, but provide a poor defense against laser technology that can cause both the shield and the cannon attacking it to explode.

POINT DEFENSE SYSTEMS

For many ships, even the presence of an energy shield isn't really guaranteed protection against incoming weapons fire. Energy weapons might be deflected by shields under most circumstances, but they're less effective against various torpedoes and missiles. Many ships, such as Battlestar-class cruisers, make use of a low-tech solution when faced with warheads, small fighters, and the like. They fire artillery shells at incoming projectiles and ships that explode nearby, spraying shrapnel into warheads and snub fighters, thus eliminating them.

RETREAT

So the saying goes: "There's nowhere to run, and no place to hide." In Space, the former is false, ad infinitum, and the latter is true nearly in proportion to the falseness of the former.

There's everywhere to run and, thusly, no place to hide.

As fun as Space battles can be when you're mopping proverbial Space floors with savvy veteran piloting, direct combat should be a last resort. No matter how many notches you've laser-etched into the hull of your fighter, it's only a matter of time before you end up another million-piece addition to the Universe's collection of Space debris.

Fear not, however. There's no shame in running away. So the other saying goes: "He who fights and runs away lives to fight another day." That one still applies. Your standard rules of cat and mouse, however, do not.

———

Still alive? Good. Consider that a job well done. Time to head back to your home planet and accept a medal. (Unless you're a wookiee. Wookiees don't get medals.)

EPISODE VIII

RETURN OF THE HERO

Rrrrrrrrrrrr-uaggghgaghggh.

—Chewbacca

CHAPTER 20

THE END...AND THE BEGINNING?

You've done it! You've won! Or, at least, you have become the face, voice, and bulge of the people and survived in the process, which is as good as victory as far are you're concerned!

And because you've won, more or less single-handedly, it's time for you to get some gorram recognition for a change. Not by your own suggestion, of course. That would appear self-aggrandizing and sour the palette of your victory musk. No, the recognition will come naturally. Portraits of your likeness will be painted in the Capitol. Tales of your heroic deeds will be whispered by hobo children around trash-barrel fires. There will be medal ceremonies and parades in your honor.

Oh, the parades. After a hard-fought war during which your government or moral constituency counted every ration bar and accounted for every photon torpedo, you're in for a shock when it comes time for them to throw a parade. Who built all these floats? Where did they get all the silklike fabric for the banners? How is it that so many tanks went unharmed in the decisive battle? Couldn't any or all of these resources have been put to better use as men froze on the front lines and entire planets were exploded? Were these items saved for the express purpose of a parade in honor of a victory that was, for a good stretch there, very much in doubt?

No matter! The parade is about many things, victory and wastefulness among them, but it's mostly about you. You, the hero. Here's how to handle yourself during your victory parade:

1. **Practice your float wave.** I'll be honest. You could smoke synthetic Space crack up there, and all the kids would be doing it in two days' time. In other words, do whatever you want from atop the float. You've earned it.

2. **Display affections publicly.** All eyes are on you. It's time to grab the prince(ss) or most desirable crew member and lay a kiss on them as they do in the crater formerly known as France. The combined thrill of victory and the public rumor mill will all but ensure his or her inability to refuse you.[1]

3. **Dress the part.** Obviously, your daily stunning Space wear is already a head-turning ensemble. But this is a formal occasion. Take your usual threads and kick it up a notch. Choose a more expensive fabric, a bigger collar, golder clasps, and tighter pants.

4. **Accept your medal with pride.** Hands back, head down, smile up. Try and forget that the medal doesn't have any *real* value, much like a Super Bowl trophy. It's a symbolic gesture.

IN THE VACUUM FOLLOWING VICTORY

Medal received, parade complete, feast consumed, Space babes satisfied. And yet, something strange is happening to you.

You look down at your mission log to find nothing on the

---

1. A firm handshake for the cohort with whom you have your differences and nearly beat to a pulp on several occasions is also acceptable here.

docket. You look up to find no three-dimensional blueprints of secret, weaponized bad-guy lairs on the holodeck.

You've achieved the galaxy-renowned heroism you set out for, so what is this unsettling lurch in your gut, this twitch in your buttock, this...feeling?

Well, Space Hero, sometimes between five-year missions and harrowing takedowns of corrupt governing entities, you will encounter something of utmost rarity. That something... is "downtime."

Yes, downtime. For in this profession, we never use the *V* word.

It feels wrong at first, and that's because it should. You're a Space Hero of action! Of adventure! A sunny beach in a three-star system or an extended stay on a pleasure planet has little appeal for a leader of your caliber (alien babes and bronze-like gleam notwithstanding). However, difficult though it may be, you should take the opportunity to relax. There is no immediate threat to the well-being of humanity and the IGFA[2]—you've earned it.

As you are surely struggling with the transition, here are some appropriate ways to spend your downtime:

- **Indulge your fandom.** Strut about in the sector of your most endearing fans. Pose for photos, accept free drinks, kiss babies, shake the tendrils of powerful politicians and entrepreneurs. The people's attitude toward Space Heroes often devolves into "What have you done for me lately?" Bask in it while you can.

- **Take the aforementioned beach vacation in a**

---

2. Intergalactic Friendship Alliance.

three-star system or on a pleasure planet. Get tan and sip fruity drinks topped with those little paper atmospheric reentry chutes.

◉ **Confess your love to the one you left behind or the one from your crew with whom you've formed an unbreakable bond through adventuredom.** Bask in the glow of continual lovemaking and the corresponding glow of your hierarchy- and regulation-free off-duty life. Plan where you'll build your home together and the names of your future children.

◉ **Mourn the death of those lost in your most recent heroic struggle.** Isolate yourself. Hit the bottle. Grow cry-for-help facial hair. Wait until your sidekick sobers you up and slaps some sense into you, or joins you.

◉ **Return home, wherever that might be.** Your rock or Space station of origin. Visit the final resting places of your family or mentor. Internally ruminate on your inability to reclaim the home you once knew. Wallow in its distance like a memory. Realize you've changed. Realize…

…That no matter how you spend your downtime, your ship is your home now. Your crew is your family. No matter how long you manage to enjoy or drink away your downtime, the feeling will well up inside you. That gut lurch—that buttock twitch. You may resist. You may hold on to the idea of the children you planned with your lover in great detail. But the call of Space will become overwhelming.

You're a Space Hero now. It is Space that made you, it is Space where you belong…and it is Space where you must eventually die.

GOOD DEATHS

Today is a good day to die.
                                    —Klingon war cry

Quick-witted, athletic, beautiful, and resilient though you may be, one can only cheat Death so many times. The immortality you've built is of the metaphorical, meta*physical* variety. But how will you die? How will you know when it's time? Will angels sing for you at heaven's doorstep? Does it tickle?[3]

One thing's for certain: your death is going to *mean* something. Space Heroes don't pass on in freak traffic accidents or at the receiving end of a villainous plot explained at length. More often than not, a Space Hero *chooses* death and has time to give heartfelt good-bye speeches in the process.

Good deaths—deaths worthy of your Space Heroism—tend to involve self-sacrifice. You want a death that is as symbolically meaningful as it is tactically beneficial. But that doesn't mean you should go around looking to off yourself. There's a fine line of planetary dust between noble sacrifice and downright suicide. Here are some scenarios that may require you to throw yourself onto your proverbial, or literal, light wakizashi.

HONORABLE COMBAT

You'll be surprised by (or, by now, numb to the idea of) how often you'll find yourself in single combat with your archnemesis or other generic villain. Typically, your wits will overcome their brute strength; your athleticism will overcome their eggheadedness; your combat roll or two-fisted club combo will overcome their advanced weaponry; your prowess with a variety of weapons

---

3. You'll have to let me know if and when you are resurrected due to fan outrage or plot contrivance or both.

will overcome their expertise with merely one; your faith in your friends will overcome their overconfidence; and your willingness to shoot first will overcome their exhausting affinity for talk. But every now and then, even a Space Hero can be bested. It's a one-on-one battle to the death, after all. Anything can happen. And what usually happens is that one of you dies.

It's not the *best* death, but there is no shame in it. If you don't get caught breaking the rules of single combat with trickery or outside help, your loss will be considered honorable, even in the eyes of your opponent. Better still, if you can parlay an honorable combat death into another one of the deaths below, your murder will be worthy of a hero of your stature.

SERVING AS A MARTYR

As a famous Space Hero, your very public and very visible gruesome death can ignite revolution. Moreover, death before the eyes of your first mate or perhaps a young mentee can serve to light the fire of heroism within them.

And though such a death, particularly at the hands of your archnemesis, qualifies as a good death, if this *is* your chosen or predestined method of biting it, there is something you should come to terms with before you go: this story isn't yours. Your Space Heroism was merely a placeholder for the *real* hero of the tale, with whom you've hopefully forged such a deep bond in a remarkably short time that she will be driven to pick up where you left off and even surpass you, so that she can accomplish what you could not.

But there is one last thing you should do before you take a laser sword to the face: leave this book somewhere the real hero will find it. He or she is gonna need it.

GOING DOWN WITH YOUR SHIP

You'll have learned by now that your ship is a crewmate unto herself. She is a mechanical reflection of your own personality, right down to warp drive malfunctions (it happens to the best of us) and a knack for squeezing out of tight spaces at the last possible moment—and not a nanosecond sooner.

So when it's finally time for your ship to retire—when she's finally taken one too many shots to the deflector shield generator or burned her engines through one too many Kessel Runs—she will do so not quietly or with a heartfelt ceremony. She will retire in a blaze of glory! She will fulfill her destiny as a ship worthy of a hero, and as a hero unto herself.

Even when your ship is shot to nothing more than a bucket of sheet metal and bolts, she always has one last bullet and one last deflector shield: herself. But it cannot shield or kamikaze alone. You'll be there, because you and your ship are one. You'll be there because, even though your ship has autopilot functionality, that shit is always the first to go when things are looking dire. You'll be there because, despite their protests, your crew can only be saved by your bullheaded display of selflessness.

A FALSE DEATH FROM WHICH YOU ARE RESURRECTED OR RECREATED IN A NEW, MORE POWERFUL FORM

Far be it for me to sap the emotional impact of a key death—*your* death—by undoing it through some absurd machination. But sometimes being dead is more of a temporary affliction than a permanent state. It's possible that you'll:

- Come back as a semitranslucent ghost, in which form you can continue providing sage advice to the actual hero.

- Leave an imprint of your consciousness or soul on another living thing, to be extracted and given a new body at a later date.

- Die merely in *spirit* (represented by horrifying physical wounds), to be resurrected as a cyborg with a limited connection to the Space Hero you once were until the emotional crux of the story offers you a chance to reclaim your soul and redeem yourself.

- Fake your own death as part of a greater tactical maneuver that gives you the upper hand on your enemy, who now believes you dead and will therefore be blindsided by your sudden but inevitable return from the grave.[4]

- See a huge spike in book sales due to your untimely but no-less-heroic demise, to live on in the hearts and minds of your loyal followers…

And though now you're now dead or repurposed as a motivational ghost or vengeful robot, you're not *really* gone. You've done what you had to do. What you always do. You've turned death into a fighting chance to live.

---

4. Has a tendency to piss off your significant other(s). Just don't expect to be welcomed back openly or be surprised to find they have been sleeping with your best friend in their wake of grief. Comes with the territory.

FINAL TRANSMISSION

"THE RETURN OF THE DEATH OF DIRK PARSEC STRIKES BACK"

Captain's Log: Thursday still. It's...hot. So very hot. I fear my sweat glands are on emergency reserves. What comes next might burn worse than the time I dozed off in an ultraviolet vitamin D bed without regulation undergarments.

Enough about my tragic death. How are you, dear reader? Chipper, I'd wager. Chest puffed and chin raised, even as you read. Good heavens, I can smell your pheromones through the time, space, and matter that divide us. You're envied and desired by people and aliens from every corner of the 'Verse. Yes, that's the look I love to see in the mirror. You really might have something there after all. From humble beginnings, you came in with a spongy yet open mind, and a flabby yet sculptable personality. Now look at you. A Space Hero!

I can't say I didn't doubt you. Not just anyone can walk in the Space boots of Dirk Parsec. But you learned from the best. I'm proud of you.

And though my death may shake you to your very core, leaving you bedridden and inconsolable for weeks on end, you'll be back. Soon enough, it'll be you out there light-swashbuckling baddies and outsmarting sentient supercomputers. Fear not for the fate of my soul, for when this godsforsaken blue star strikes me down, I shall become more powerful than you can possibly imagine. I will transcend my human form, impressive though it may already be, and ascend into legend...

That's about it for me, then...I can feel my...lungs...clinging to what little oxygen remains in the ship. Damn you, you cerulean devil! Damn you, you burning blue bastard!

Yes, it's nearly time now...My skin is burning. Oh, for all that is holy! This is the end!

Heed my words, Space Hero! Honor your mentor! Avenge my death by...by telling these burning balls of fission reactions who *truly* rules Space!

Back to...back to the...

...

stars...

...

...

...

...

..

.

All right, Agulor, that should do it. Kill the dictation, override the escape-pod data logs, and jettison the black box. With any luck, some poor sap will find the contents of my collected wisdom. The death of Dirk Parsec will spread grief to the far reaches of the galaxy. And in their mourning, billions will buy my last words to see what knowledge is there to be gleaned from the greatest Space Hero ever to soar these icy, black skies.

Hell, who knows? We may even trick Flabba Kohn into believing I really am dead.

On the double, then, Commander. Let's hit the escape pod. Your captain needs a drink, and he knows the perfect little backroom poker club near the Crab Nebula. I'm pretty sure "Dr. Derk Parschleck" still has some credit there.

ACKNOWLEDGMENTS

Like a preternaturally competent Space Hero with amazing feathered hair, books have a tendency to seem as though they leap into existence, ready to pull the ears off a gundark. But like Space Heroes, books wouldn't be possible without the hard work and influence of many more people than whoever's face is cast in bronze on your newly conquered planet.

Chief among those whose invaluable work made this book what it is, and way better than what it was, are Sourcebooks senior editor Stephanie Bowen and assistant editor Jenna Skwarek, who brought not only a great eye for cutting (it needed so much cutting), but who also offered research the authors, uh, neglected to think of, and science fiction insights the authors overlooked. Their knowledge and commitment to this project were nothing short of glorious, and they deserve a medal ceremony for their efforts. (The wookiee didn't help at all, though, and deserves nothing.)

Some special thanks to Daniel Villenueve, whose patience persevered through an insufferable amount of "creative input" by the authors and whose pensaber brought to life Dirk Parsec, his beautiful hair, and his merry band of Space Misfits; to Isaiah M. Johnson, whose fantastic interior and cover designs tricked you into buying this book; and to Rachel Kahn who, despite her name, was an ally, and without whom we would have capitalized damn near every word in this Book.

Also indispensable to the rise of Dirk Parsec and his incredible wisdom rendered into book form was Brandi Bowles. As a literary agent, Brandi is second to none, and helped craft the early idea of this project from little more than a punny title and a vague notion of helmet hair prevention into something that could actually be written and, maybe, make people laugh. She remains the desert hermit mentor to the authors' confused and whiny moisture farm boys.

This book also would have been impossible if not for the support of Amanda and Caitlin, who put up with not only countless hours of hogging the TV to rewatch *A New Hope* for the twelfth time (for "research"), but who also are early readers and editors, sounding boards and idea shapers, and unwitting joke testers. When the hyperdrive motivator of the authors' lives goes finicky and they're left bashing it with a wrench and howling shrilly, Amanda and Caitlin are known to grab a hydrospanner and keep things flying. Maybe this analogy is starting to break down. We love you both.

Finally, there have been many a Saturday matinee and Sunday hour spent with Space crews and Star heroes that led directly to the creation of this book and were made possible specifically by a few dads and moms. Mike's basketball card collecting gave way to a child's comic book collecting, and he paid for it dearly with countless trips to conventions and far-flung comic shops. Lanell was supportive and curious of geeky ideas, even when she didn't understand them (which was always), and always fostered the channeling of those ideas into creative endeavors.

Years of *Star Trek* episodes consumed on the couch beside Carl were more formative then he could probably ever know, and far more so have been the many conversations over the years about science, science fiction, stories, and possible incredible futures. Rosie's willingness to provide books whenever they were desired, as well as to put up with sci-fi movie rentals, and Greg's shared enthusiasm for great (and dumb) movies and TV shows were similarly both incredible and essential to this book and much more. Their tireless, unwavering support has made so many amazing things possible for these authors, and there may not be enough gratitude in the Universe for all they've done. The authors will see about starting with some bronze statues on distant planets, however.

ABOUT THE AUTHORS

Nick Hurwitch plans, in the unlikely event that he dies, to sidestep astronaut training and have his body jettisoned into Space. Until that time, he plans to write as many screenplays, video games, and books as possible, and asks that you donate generously to the Nick Hurwitch Space Burial Fund. He lives in Los Angeles with his almost-wife, Amanda.

Photo by Nick Ahrens

Phil Hornshaw is a would-be Space cadet who failed to win a trip to Space Camp in his youth, despite entering a number of cereal box sweepstakes, and so chose to pursue TV watching, video-game playing, and writing instead. When he's not trying to find out what makes proton torpedoes more protony than regular torpedoes, he works as a tech and video games journalist. He currently lives in Los Angeles with his wife, Caitlin.

Photo by Keith Owens